How to Learn and Practice Science

A. R. Prasanna

How to Learn and Practice Science

 Springer

A. R. Prasanna
Ahmedabad, Gujarat, India

ISBN 978-3-031-14513-1   ISBN 978-3-031-14514-8   (eBook)
https://doi.org/10.1007/978-3-031-14514-8

This Springer imprint is published by the registered company Springer Nature Switzerland AG
The registered company address is: Gewerbestrasse 11, 6330 Cham, Switzerland

Dedication
Entaro Mahanubhavulu, Andariiku
Vandanamu
(Saint Thyagaraja)
(Salutations to all the Great Souls, Learned
and Wise)

Preface and Acknowledgments

My interest in science, which started in Class 8, got boosted when I had my first occasion to meet with Professor Sir C.V. Raman, at his institute in Bangalore in 1957 when I was in Class 11 (Matriculation). To my fortune, I had a couple of more meetings with him as I had registered to attend his annual lectures on Gandhi Jayanti on October 2 every year. One of those (in 1961) was the most significant for me, as arriving early at his institute, I had the rare privilege of walking along with the Professor in his lovely rose garden listening to him about science and how it should be practiced. Oh, what a great person he was! One of his lectures on "Light, Colour and Vision" is still fresh in my memory.

As I chose to specialize in a different area of theoretical physics and left Bangalore after graduation, I had no further occasion to meet him again, but the early memories are everlasting. Though I have been a practicing student of science all my life, I had learned how important it is to popularize science and scientific temper through interactions with students and society in general. My own very first popular lecture was delivered in the then British Council Library in Madras (now Chennai) in 1965, while I was a research student at the Institute of Mathematical Sciences, popularly known as Matscience. I do enjoy giving popular talks mostly on the topic of "our universe," as also the methods of learning and doing science. I do believe that science is to be learned and not taught.

With this philosophy, I have engaged school teachers on different occasions urging them to inculcate in their students the spirit of learning science. During the last three years, staying home (due to lockdowns and pandemics) I started putting together some of the material I had used for such workshops and lectures in the form of essays, and this little book is a result of that venture.

The plan of the book is geared toward both students and teachers with the earlier chapters concentrating more on students and later ones initiating the process for teachers (communicators). Apart from these two groups, the book may also be of interest to anyone interested in science and likes to develop scientific understanding and scientific temper.

I sincerely believe that children are by birth curious about everything that they see and experience. As adults, we need to channelize them to think, ponder, and arrive at understandable conclusions which they can appreciate and communicate to others whenever required or questioned.

The book starts with the topic "what is science?" The basic aim of science is to understand the various happenings around us and how our ancient generations tried to interpret attributing reasons and causes to effects observed rather than attributing to the supernatural. Though initially, the attempts were to make life easier with some innovative gadgets, the more inquisitive among them turned their attention to understanding the cosmic order. Though not exactly, about six to seven thousand years ago the humans in different parts of the world must have started wondering about questions like where, when, and how did the world come about.

Is science only esoteric or can one see it operating in daily life? It is something we all see and experience in day-to-day routines if only we stop and question after the action as to why did I do it? Once observed and analyzed the learned concepts can lead to events beyond and help in trying to follow the cosmic order. While we are very adept at using the technological outcomes of science, one needs to think of the hidden science behind the technological innovations. One should indeed feel obligated to those pioneers who spent their facilities and time to look for applications of scientific knowledge and improvise gadgets that we use in daily life.

Is there a methodology for learning and doing science? Immanuel Kant's explanation of the philosophy of getting new knowledge both through reasoning and through new observations underlined an established and well-practiced methodology in the world of science. Discussing the results obtained with others and verifying them thoroughly are very important features of scientific methodology. Human knowledge initiates the thinking process. The reason for the methods of science being difficult is because any result obtained either by experiment or by observation has to be repeatable. Whereas Plato's school in Greece initiated the process of trying to explain the celestial motion of stars and planets, their approach of a geocentric system was replaced with the helio-centric system developed by Copernicus. The a priori (*apriori*) (pre-knowledge) and the a posteriori (*aposteriori*) (outer knowledge) concepts as put forward by Kant allowed equal emphasis on both the relationship point of view and the empirical point of view. Though the methodology cannot be compartmentalized, the theoretical and the experimental viewpoints are very important both in establishing earlier learned knowledge and in confirming its extension.

How should one communicate science? Apart from audio and video modes which are normal, a more important aspect of conveying the concepts and experiences needs to be put in a form that is easily understood by the receiver. One should bear in mind the fact that just giving a whole lot of information does not convey science. Successful communication is that which brings in the joy of understanding to the receiver. As science deals with the knowledge of nature, its communication could and should be in and with all possible forms of perceptions. Once a phenomenon is described or a law stated, it should produce the same effect or give the same result irrespective of when or where the experiment is conducted or observation is made and who does it. This would indeed initiate the question: does thinking need a language? Further one should demarcate the differences between information and knowledge.

Mathematics as a common base for science has and is playing a fundamental role in quantifying the ideas expressed as well as in expressing logically the cause–effect relations in theories and models. Puzzles and riddles often pose a thinking mental pastime and sharpen the mind through a quest for solutions. Through the usage of coordinate systems, mathematics provides an apparatus - the language of numbers which between different observers are linked by finite and reciprocal transformations. There have been concrete examples where discoveries and theories are developed by adopting new mathematics to advance old theories. One ought to remember that though all mathematical developments have their psychological roots in practical requirements, it invariably gains momentum in themselves and transcends the confines of immediate utility as expressed by distinguished mathematicians. Mathematics plays important role in establishing working relations between different disciplines as may be seen in the development of subjects like biomathematics, computer simulations, and information technology. With all its seriousness, mathematics can be fun too and improves the skill of any individual who spends time, solving puzzles as a pastime.

Science is an integrated adventure as demonstrated by practitioners of earlier eras. Unfortunately, the new advances have rendered it to severe specializations which often may make a practitioner lose track of the associated concepts. This must be looked into, and particularly at the elementary level, one should communicate science in an integrated way showing the young minds the applicability of one discipline in the other associated disciplines. This requires a thematic development of teaching aids and models. Topics like motion, work, force and energy, structure, and states of matter are all examples of scientific concepts which can be communicated in an interdependent and integrated manner. While super specializations cannot be avoided, they have their role only at the research level and not for communicating (aiding to learn) at the elementary level.

Experience and experimentation are hallmarks of learning and practicing science. Man's early experiences from childhood are indeed the basis for an inquiring mind which grows into the stage of experimenting to know more about things around us. While trying out simple experiments during adolescence may be considered adventurous, pursuing that approach to learn more as one grows into adulthood shows clearly a scientific bent of mind. As one learns more and advances into research level and keeps the habit of questioning, trying to figure out and finally get to work in a laboratory is not just feeding the curiosity but setting up oneself for a scientific endeavor. When Galileo kept on rolling balls down an inclined plane with varying inclinations, he was seriously trying to find out the path which took the least time to roll down. Imagine the surprise he had when he realized that the path was not a straight line but along a curve (the Brachistochrone), and this truth established a whole new way of thinking. Most of our day-to-day experiences if they occur time and again result in the realization of some universal truth and a real scientific mind will then work on the verification of such facts by experiment or observation. Several times it could be an accidental discovery but still needs verification through controlled experiments.

Doing accurate observations of events is as important as doing laboratory experiments, and this is particularly important when studying the cosmos and the motion of cosmic bodies. The accurate observations of Tycho Brahe and his collection of data helped Kepler work out the orbits of planets and the laws that govern them. The success of the Newtonian theory of gravity is mainly because using Newtonian mechanics one could verify Kepler's empirical laws. Sometimes observations could come accidentally, which is referred to as "serendipity." The earliest perhaps was the discovery of fire both as a facility and an armor against wild animals. In life sciences, the discovery of penicillin was one such, whereas in astronomy the discovery of the radio universe, which though accidental, required very keen observations for confirmation. The book tries to illustrate several such findings.

Till the advent of the twentieth century, scientists working on research problems did so purely for their academic satisfaction and enjoyment without worrying about returns. Post-industrial revolution, doing research turned into a profession, initially to develop "warfare" but later to look for aids to benefit the society through industrial applications of science-based technology. Post-1960s people slowly realized that apart from aspiring for degrees in engineering and medicine which were considered lucrative professions, studying pure science and going for research in its pure and applied aspects could also be considered as a profession. Have we understood scientific research as a full-time profession like any other though maybe less rewarding? Unfortunately, a very small percentage of society realizes this. The most important aspect to be appreciated is the fact that scientific research is not a 9 to 5 job! It is the only profession where one gets paid for doing what he or she loves to do and moves a step ahead in understanding nature. One must say that over the last two decades, the financial remuneration for science professionals has been far better than what it used to be and chances of getting a good position depend entirely on one's merit and dedication to work.

Finally, all the activities of human beings are linked to the society they live in. As such it is necessary to look at the way society is interacting with those practicing science both as an intellectual adventure and as a profession. A glance at the history of science reveals that knowledge advancement and understanding nature in its various forms have only been possible due to the untiring and dedicated efforts of men and women analyzing what they observed and what they experienced. Early Greeks like Plato and Aristotle looked for patterns in the behavior of things both living and non-living. It was conveyed that nothing should be taken for granted and whatever knowledge one had from the earlier generations should only act as stepping stones for a new adventure. Scientific research is thus a continuous process with a firm beginning but never-ending. Unfortunately, the society which includes scientists themselves does not seem to have a real understanding of what scientific research is and the number of efforts that practitioners of science put in toward understanding how nature works or why is it unique. Of course, the valid reason could be the fact that very few scientists take time off, if at all, to communicate to the general public about science, its values, and its contributions to the growth of human society. In essence, this requires developing a scientific temper for which both the practitioners and their communities should work synergistically so that the entire human society can overcome the ignorance of superstitions and learn to look for proofs and logical reasons for all the cause and effect relationships.

In my life of science over the last sixty years, I have taken every opportunity I got to interact with people of different expertise and experience (teachers, scientists, colleagues, and students) and learned whatever I could, which in turn could have influenced my views while writing this book. I am very grateful to all of them. More directly, the reviewers of the drafts of the manuscript at different stages gave some good and helpful suggestions which I have tried and incorporated to whatever extent possible, and I am very thankful to them. I am thankful to Prof. A. K. Singhvi for his reading of an earlier draft manuscript and for making a few helpful comments. I would like to express my appreciation to my two young friends Mr. Sonam Bhatt and Ms. Kavya Shah, who read through some of the chapters in the earlier version and indicated a few difficult terminologies from a student's point of view, which I have tried to clarify. I must say that the final version in your hand is the result of very useful suggestions and comments by Dr. Ramon Khanna, the editor, and his colleagues at Springer publications. I am very thankful to them. After I had finished my first draft, my attention was drawn by Dr. Khanna to the books *The Rise of Science* by Shaver (2018) and *Technology and the Growth of Civilisation,* by Genta and Riberi (2019). For those particularly interested in finding out more examples on aspects of mathematics as a basis of science, and the interdependence of different disciplines of science and technology for successful research, I consider these books very useful.

I would like to express my love and appreciation for the continued support and understanding received from my family, wife Shanti, children Kartik and Tanusri and their spouses Bidisha and Fahim, and my lovely grandchildren Ananya, Anoushka, Sophia, and Rehan, during consolidating this work (in the difficult times of the pandemic). I would like to thank the concerned authorities in the Physical Research Laboratory for the library and computer facilities provided. Despite all my efforts, there could still be scope for improvement and I will be happy to hear from the readers, their views, critiques, and suggestions. I may be always contacted through email at aragam@gmail.com

Ahmedabad, India   A. R. Prasanna

Contents

Chapter 1
What Is Science?

Introduction

Science as understood and practiced is the process of learning about Nature through observing and questioning events happening around us. In this chapter, starting with the definition of science, we will consider the early beginnings and purpose of science. In particular, we will examine how natural events trigger our curiosity as we seek reasons for seemingly inexplicable occurrences. Historically, our beliefs and practices were often based on superstition, but once the fear of Nature was overcome, humans started to question how and why events occurred. The drive to make life easier led gradually to innovation and the advent of technology. The more inquisitive among them turned their attention beyond the Earth to seek explanations for the cosmic order. In this way, the urge to learn about the reasons for different natural phenomena sets us on the path of seeking knowledge for its own sake.

We live in a Universe full of matter and energy that spans large scales of space and time. But how did it all begin? How do various processes in the Universe operate, and what makes them interconnected across the various scales of space and time? Understanding the patterns and the fundamental relationships that govern the Universe has been arguably the greatest adventure that humanity has engaged with. Though the human race has a history of about three million years, it appears that the homo sapiens, the direct ancestors of present-day humans appeared only about two hundred thousand years ago but started the migration to other parts of the world only about fifty thousand years ago. Even then the actual settlements of humans seem to have started only about ten thousand years ago when they discovered agriculture, domesticated animals as also learned the idea of getting metals from their ores (Shaver, 2018; Genta & Reberi, 2019). Though the biological evolution of the human species happened throughout in different steps and periods lasting millions of years the mainline leading to the final step of modern man (woman) came with the advancement of the brain. This evolution seems to have happened not only with the increase in mass but also increase in brain size with the cerebral cortex getting more complex permitting an increase in the brain surface without increasing the volume. It is indeed remarkable that the evolutionary process leads to changes in the anatomy of the pharynx capable of producing different varieties of sounds as well as developing the capacity to walk in an erect position (Genta & Riberi, 2019). After about another few thousand years the human mind seems to have begun to wonder as to the origins of the world around leading to questions like what happened at the beginning, when, and where did it take place? Why did it occur, and how did it happen? Questions that formed a cognitive revolution of the humans.

At the most fundamental level, science is nothing more or less than pondering on these questions and attempting to answer them. In other words, science is both knowledge itself—Scientia in medieval Latin- and the process of acquiring it—Scintificus. More formally it has been defined as the pursuit and application of knowledge and understanding of the natural and social world following a systematic methodology based on evidence (UKSC09).

Early Times

While discussing science as a discipline one may not be able to exactly mark the period of the evolutionary history of humans concerning the development of the thinking process. Anthropologists are still working and analyzing available data in this regard and that can indeed be an independent line of research for understanding human history. However, it is perhaps not unreasonable to assume that once the survival and safety issues were taken care of, some among the early humans purely out of curiosity could have used some time looking around observing and reasoning for the happenings around them. Also, it could be the fear of the unknown and the urge for survival against furies of nature that lead the human race to worship Nature. Slowly they must have realized the critical role of Nature for the very existence and sustenance of life. The 'human mind' could then have ventured into the arena of how and why of things which could have indicated the cause and effect relationships of events leading to the beginnings of scientific thinking. For more details on these aspects of science and technology one may refer to the earlier quoted references, Shaver (2018) and Genta and Riberi (2019).

It is known that curiosity leads to discoveries and inventions. The history is replete with examples suggesting that many times accidents led to discoveries; the earliest and most significant being the judicious utilization of fire which altogether modified the lifestyle of humans. For several centuries the only way for humans to interact with nature was through their sensory organs—viz. Eye—Light, Ear—Sound, Nose—Smell, Tongue—Taste, and Skin—Touch (hot or cold, hard or soft) as the probes to experience varied manifestations of Nature. The fear of the unexpected and the urge to gain supremacy over others generated superstitions-blind beliefs passed on from generation to generation. Only with time, these beliefs led to a compromise on reasoned and evidence-based thinking. The practice of superstitions was carried over for a long time even though superstitions have no methodology of analysis and are the results of blindly following the beliefs of one's ancestors. It is still prevalent in most cultures and religions and sometimes can be the cause of misunderstandings and unhappiness. Science questions superstitions. It is only when some logical and fearless minds were willing to critically analyze the occurrences around them and establish the real basis using the methods of reason that saw the progress of civilization and forward movement for scientific thinking.

Let us consider some experiences that the human race would have had and the kind of reasoning it would have gone through to discover facts about Nature and the laws that govern it. The earliest of the experiences which were apparent to humans should have been the change of day and night, associated with the rising and setting of the Sun. The first reaction obviously would have been a sense of reverence towards the Sun leading to the practice of worship. Varied records suggest that the human race all over the world has worshipped Sun as being the eternal power that initiated and then continues to sustain life on Earth. These practices have continued till now in all cultural backgrounds and regional/religious identities. Such reverence to the Sun is buttressed by science, which also teaches that the Sun is the prime source of energy for the Earth. The pictorial depiction of the Sun in different cultures may be seen in the following painting given below (Fig. 1.1):

Fig. 1.1 May we contemplate upon Bhaskara, the shining one. Let us meditate upon the one that produces great illumination. May Aditya illumine us. Thaittareya Upanishat. Credit: Chandranath Acharya, J Nehru Planetarium, Bengaluru

The belief that the Sun is the key to the sustenance of life on earth is true scientifically. However, it is only a small fraction of the society that is aware of the science of what the Sun is and the processes that make it the perpetual source of energy. Despite the limited understanding, a large percentage of the population (thinking that seeing the eclipse causes ill effects) stay indoors during an eclipse and therefore miss out on experiencing the immensity of celestial beauty that unfolds itself during a total Solar eclipse (Fig. 1.2a, b).

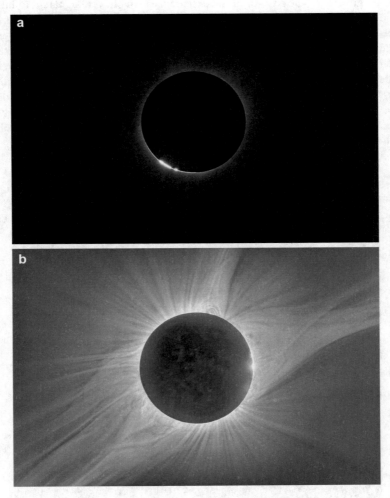

Fig. 1.2 (**a**) Diamond Ring. (**b**) Solar Carona (courtesy: NASA picture gallery) (**a**) https://www.nasa.gov/image-feature/the-bailys-beads-effect-during-the-2017-total-solar-eclipse. Photo Credit: NASA/Aubrey Gemignani. (**b**) https://www.nasa.gov/sites/default/files/thumbnails/image/3654 9747932_2ba72f7631_o.jpg. Credit: ESO/P. Horálek/Solar Wind Sherpas project

Luckily during the last two decades, many of the science popularisation groups are making continued efforts to avail the general public with visual aids to view the total eclipse as well as medical advisories to instil confidence against misguided beliefs.

When confronted with the questions of natural phenomena the early humans sought answers based on what they thought they saw and as a first reaction ascribed these to the work of the unknown and unseen agencies (the spirits!). Such simplistic suggestions found common usage. Further, it was propounded that like people, these unknown and unseen agencies too had their moods as being- happy, sad, angry or jealous—depending upon which, they caused natural effects—good ones like rains on parched earth, or the bad ones like floods, disease, and earthquakes. This misguided notion continued for a long time and most views of the world were borne out of fear or fear-induced reverence. People used incantations or potions or sacrifices to appease these hidden entities (spirits and gods) or to cure illness or misfortune and miseries. It is ironic that at the same time people also developed tools to try and control the world around them. Thus during the old stone age (circa 2.4 million years ago), they tended materials to make weapons for hunting and by the new stone age (about 10,000 years ago) the humans began understanding nature, like for example how plants grew, a knowledge that paved way for agriculture as a systematic practice to grow more food.

Technology Versus Science

The common adage, necessity is the mother of invention is borne out in the evolution of mankind. But, strictly speaking, these inventions were not science, only technology. Though many times used interchangeably, science and technology are different. Science is the pursuit of basic knowledge whereas technology provides usable products, based on scientific principles. It is however interesting to find that historically speaking technology based on intuition preceded scientific studies. Nowadays the advancement of science and technology occurs synergistically—with the one supporting the other in a bootstrap process.

In the beginning, almost all efforts to make tools developed from stones, wood, and metals, were mainly to fight the wild animals and for their use in agriculture. Great river basins like Nile, Euphrates, Tigris, Indus and the Yellow River were the sites where the agricultural civilizations got initiated along with the domestication of animals mostly around the Neolithic period (Genta & Riberi, 2019). Around 3000 BC Sumerians seem to have domesticated animals which were used to plough the land. These communities also built ships for trade across seas and had later discovered how to make bronze from copper and tin. Some historical factors suggest that agriculture was practiced in ancient India as early as 9000–8000 BCE, along with the domestication of sheep and goats, and the planting of grains such as wheat and barley. Ancient Egyptians were not far behind in these developments, and both these civilizations had developed simple numeric systems for keeping accounts. All these civilizations appear to have made technological developments even in those times largely based on observation, experience, and trials which is an indication of the initiation of thinking process.

Initiation of Scientific Inquiry

It is generally surmised that the development of science as practiced in the last few centuries began with the Greeks of the Macedonian peninsula. Though they had elaborate mythology with spirits and gods, they were perhaps the first to seek answers for various naturally occurring phenomena through their philosophy and possible deductions related to experiences. Curiosity and a possible maturity in thinking seem to have led them to make a transition from reliance on myth to the search for a logical explanation—in other words, what we would today refer to as scientific principles. They sought to find general patterns in nature and made an effort to determine their Order! It is worth noting that many of the Greek philosophers relied on subjective thoughts and their intellectual efforts comprised very little of experimentations and observation. The same may be said about the traditional Indian Vedic science too, which also seems to have been dated to the third millennium BCE (Kak, 2005). The period 1500 BCE to about 600 BCE, the time Aryans from central Europe migrated to the northern parts of India is referred to as the Vedic age. Though there was no material evidence available that could be dated, several ideas covering the areas of mathematics, astronomy, and medical practices seem to have existed in this period. The post-Vedic period (600 BC to 900 AD) is considered to be the golden age of ancient Indian science and understanding of traditional values (BSS, 2017).

Summarising the ideas mentioned, one can define Science as follows:

Science is the knowledge of natural phenomena and their consequences. It is an evolutionary process comprising objective observation, seeking evidence, attempting experiments to test the hypotheses. It is not just a collection and collation of known facts. As we will see in the following, science is a philosophy derived out of experience, innovation, and verification or validation.

The Purpose of Science

Having defined what science is, let us understand the purpose of science. Since the time the early man realized that fire can be both useful and harmful his/her mental abilities (supposed to be the highest in the animal world) rationalized through reasoning the advantages of his/her action which at times were simple reflexes to natural causes. One could indeed say that this could have been the beginnings of the usage of higher mental capacity, the gift humankind has from nature. Growth of mental capacity in animals like dogs, monkeys, and dolphins as shown by their reactions to some of our actions and keen observation of the behaviours of birds and insects too show a relatively progressive understanding in their actions and reactions to several natural events which do indicate their capacity to gather and interpret many of the actions of other creatures. It is obligatory while doing science to look for Nature's providence to all forms of life on earth which has been the basis for progress of human civilisation.

As evolutionary biologists seem to indicate, Man began evolving from his very early ancestors some twenty million years ago and possibly developed the first thinking mind in an animal a few million years ago. This originality and ingenuity enabled him/her to reign over other animals due to his/her ability to use hands and invent gadgets/tools to help them against adversities; natural or from other animals (including others of the same species!). This capacity was the basis for a technology whose aim has been to enhance the comforts of living and develop facilities and gadgets to gain supremacy over all living beings and even over nature!

The same human mind which has evolved over at least a million years to a mature state did not remain satisfied with physical comforts of the body, security of food, water, and energy, and just safety of the self only. It had the stimuli and the desire to learn about the unknown and understand things that appear commonplace to many but follow specific patterns governed by the intricacies of nature and its laws. It is only but a few who engaged themselves in delving deeper into the questions on the causes for natural order. Such minds (called natural philosophers/scientists), created new paradigms and found answers through hypothesis, logic, and experimentation—which are the due process of Science.

Stated simply, the purpose of science is to obtain a complete understanding of the fundamental laws that govern the Universe. In its quest for enhanced understanding of nature, one used the existing knowledge to develop new hypotheses and invented new technologies to assist more advanced exploration. This process and the synergistic relation between science and technology has continued to the present day and is happening at an ever-accelerating pace.

Some Landmark Science-Based Technologies

One of the early inventions of humans that made their life easier was the wheel. It is not exactly known when the wheel was used in the form it is known as there is no evidence of any scientific theory for its invention. Reasonable guess indicates that for the early human, who might have observed a round block of log rolling easily over a slope could have been the precursor of a wheel. Later improvements in the design of wheels and their possible applications in different contexts could certainly be categorized as an outcome of scientific thinking. The same trend continued as human ingenuity progressed in associating events and causes. The intertwined role of science and technology became increasingly clear such that every new technology assisted the development of new ideas in science; and developments in basic science helped the advancement of new technology. During the past five hundred years, the technological revolution and the progress in science have been at an ever-increasing pace. As Bertrand Russel has said, 'in the beliefs of educated men, science has existed for about 300 years while as a source of economic utility, for about 150 years and in this brief period it has proved itself an incredibly powerful revolutionary force'. Despite technological innovations occurring at an accelerated pace, there are still several fundamental problems in science that need further explanation and understanding.

It is important to realize that the new technology is derived from the understanding of basic scientific facts concerning a) the properties of matter under different natural and laboratory created controlled environments and, b) through the interactions between the physical, chemical, and biological forces that exist in various manifestations of nature.

As it has been aptly said, *today's science is tomorrow's technology*. In the following, some interesting developments are narrated to substantiate this feature. One of the earliest may be the discovery of vacuum and the barometer by Toricelli (1608–1647) a student of Galileo. This led Robert Boyle (1627–1697) find a relationship between the pressure and volume leading to Boyle's law, which helped understand several aspects of the physics of matter and along with 'Charles law' (relating the volume and temperature) is fundamental for the study of the kinetic theory of gases and the mechanics of fluids. Toricelli's work and his experiments with Otto Von Guericke on creating a vacuum seems to have laid the foundations for understanding the concept of atmospheric pressure which later paved the way for the invention of the steam engine by James Watt. This needed the idea of using the steam pressure to push the piston as well as having a new separate condenser (to cool the steam back to water) that increased the efficiency of the engine. The application of steam engines for generating power for machines was the beginning of the industrial revolution. However, this soon changed with the usage of coal and fossil fuels for generating power and this change automatically lead the steam power to be used for locomotion and navigation (locomotives and water boats). The same work on measuring the atmospheric pressure contributed to the later developments of high flying balloons which eventually lead to the flying machines and the air transportation.

It was a phenomenal change that was brought in by the discovery of electricity without which it is difficult for people to live comfortably even for a few hours. In today's world our dependence on the electric power ranges from the use of normal household gadgets to running industrial units of every type. We must owe our gratitude to Hans Oersted (who discovered the magnetic effect of current) and Michel Faraday (who discovered the electromagnetic induction) for the multiple comforts provided through the use of electricity that we enjoy today. In this context, an interesting anecdote runs as follows. When in 1821, Faraday discovered the dynamo (electromagnetic induction) he wanted to demonstrate this to the public. He arranged the experiment and invited the public along with the mayor of the town for a demonstration. When the setup was ready and he started the motor and the lamp in the circuit got illuminated, the public clapped but the mayor looking at Faraday appears to have commented wryly, 'Oh what use is this?' Faraday seems to have responded politely, "Sir, by next year you will be taxing this". How true! The entire industrial revolution stood on the power run machines using steam and electricity. The technological inventions then followed with the discovery of electromagnetic induction (motor) and the generator converting mechanical energy to electrical energy (dynamo) and vice versa. Various types of motors and generators, working both on direct and alternating currents were developed along with transformers that helped in transporting electricity over far distances safely. Further with future inventions of electric bulb, Morse code, the telephone, and phonograph, electricity got a prime position in human life both for work and entertainment. It is also important to remember that Faraday's contributions to science went beyond the study of electromagnetics as he is credited with the discovery of some carbon compounds (1820), and chlorine (1825) apart from several others. Today with the new appreciation of climate change and associated carbon foot prints, countries are encouraging the public to use electric cars and busses for transportation. This will be good for saving fossil fuel as also to reduce pollution.

Often unnoticed, the most important technique of harnessing Nitrogen the gas that makes up a large percentage of air, by the German chemists F. Haber and C. Bosch for making the fertiliser gas Ammonia, (by combining with Hydrogen) in 1910 can be considered as a very important step in feeding the hungry mouths all over the world. According to experts a large portion of nitrogen in our bodies almost 80% comes from the Haber-Bosch process. Some even consider this as one single reason for population increase in the last one hundred years. Nitrogen seems to play a very important role in the biochemistry of all living creatures (Lorch, 2015).

An important discovery made in 1888 but had to wait for more than eighty years before its importance was realised was the discovery of liquid crystals. Everyone is familiar with the usual three states of matter, but the possible existence of a substance having liquid phase and a crystalline phase at different temperatures was indeed a novelty. Austrian botanical physicist F. Reinitzer working on derivatives of cholesterol had noticed that 'cholesteryl benzonate' displayed two melting points, melting into a cloudy liquid at 145.5 °C and melting further at 178.5 °C when the cloudy liquid becomes clear. He also noticed that the phenomenon was reversible. He had further noticed that apart from having two melting points it reflects circularly polarised light (light wave in which the electric field vector rotates around the axis in a plane perpendicular to the propagating direction) and also rotates the plane of polarisation (Fig. 1.3) circular polarisation.[1]

Further research in this area led to nematic liquid crystal (transparent or translucent liquid which can change the polarisation of a passing light wave) at 125 °C and produced a regular pattern which subsequently led to the possible use of liquid crystal panels as display boards instead of cathode ray tubes. However this

---

[1] The electric field vectors of a traveling circularly polarized electromagnetic wave. This wave is right-circularly-polarized, since the direction of rotation of the vector is related by the right-hand rule to the direction the wave is moving; credit: Wikipedia contributors. "Circular polarization." *Wikipedia, The Free Encyclopedia*. Wikipedia, The Free Encyclopedia, 11 Jan. 2022. Web. 8 Feb. 22.

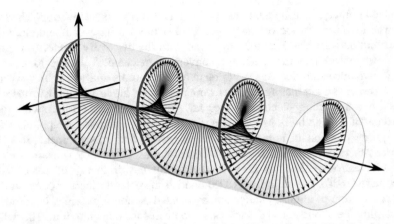

Fig. 1.3 Circular polarisation: credit https://commons.wikimedia.org/wiki/File:Circular.Polariza
tion.Circularly.Polarized.Light_Without.Components_Right.Handed.svg.   Dave3457,   Public
domain, via Wikimedia Commons

application had to wait till 1966, when Goldmacher and Castellano discovered a
newer method to produce liquid crystals at room temperature and subsequently the
study of liquid crystals in all its possible phases and orientations turned into a huge
industry among the condensed matter physicists and chemists. Research on biological liquid crystals and mineral liquid crystals have shown immense possibilities of
applications in a wide variety of situations. Like in electro optical devices in
hyperspectral imaging and thermotropic chiral liquid crystals as thermometers as
their pitch varies with temperature and liquid crystal lenses in adaptive optics as they
converge or diverge the incident light by adjusting the refractive index of the layer.
Liquid crystal films have revolutionised the world of technology as they are used in
almost all possible daily utilities like mobile phones, televisions and computers.[2]

The discovery of X-rays by Wilhelm Röntgen in 1895 can be considered to be
accidental, which occurred during his studies of the effect of cathode rays on metals.
Finding that the newly found radiation traversed through barriers and produced
images on screens he realized the possible applications of this radiation and accordingly drew the attention of other scientists to the discovery. The first-ever X-ray
picture of human anatomy was that of the hand of his wife showing clearly the
structure of finger bones including the wedding band. The significance of this
discovery (which got the Nobel prize in 1901) and the possibility of its applications
is very well known particularly in medical and mineralogical studies. Deductions
from X-ray crystallography, made possible through the study of X-ray diffraction by
W. H. Bragg (Nobel prize 1915), played a very fundamental role in the discovery of
the structure of DNA. In the medical profession radiology is a very important branch

---

[2]Material for this discussion is based on the aspects of liquid crystal from the source https://en.
wikipedia.org/wiki/ Liquid crystals.

for diagnosing patients with different types of internal diseases. Using CAT (Computerized Axial Tomography) scan showing the X-ray images of the body parts is almost a routine practice in medicine today. X-ray fluorescence is a very important technique that helps the elemental analysis of samples by using X-ray diffraction methods. One of the most used tests for security checking in air ports and other sensitive areas is through low-energy X-rays. Thus a large number of day-to-day applications in different sectors have resulted through the discovery of X-rays also called Röntgen rays.

Similar to Röntgen's discovery of X-rays, was the discovery of the most common plastic material-the polythene, which is considered as an accidental discovery that happened twice. The first was in 1898 when the German chemist Hans von Pechmann noticed a waxy substance at the bottom of his test tubes while working on an entirely different project. Detailed studies of this substance by him and his co-workers revealed the fact that it was made of long chain molecules which was named polymethalene. But unfortunately no further studies were made by them. Fortunately in 1933 an entirely different group of chemists working on the production of plastic using a different method but under high pressure noticed a similar waxy substance as was seen by von Pechmann earlier. This was turned into a practical method of producing plastic resulting in what is today called the polyethylene or polythene. In the last few decades the plastic (polythene) industry has grown so much that there is an estimate of about 100 million tonnes of polythene resins being produced annually. The types of products made out of this substance range from food wraps to detergent bottles, visors and hard hats to trash bags and fuel tanks for automobiles. *Despite the cheap availability and being lightweight for transport, polyethylene has a big disadvantage. As a waste material it cannot be completely destroyed as it is not biodegradable.* This can cause a very hard impact on the environment particularly because their debris, laced with chemicals ingested by sea and wild animals can get poisoned. Floating plastic waste which can last for centuries and more, can provide transportation for harmful species that can disrupt habitats through ground water and landfills carrying harmful chemicals to all life forms (Knoblauch, 2020).

The discovery of the atomic and molecular structures of elements and compounds led to exceptional innovations in material, medical, and pharmaceutical industries and has helped society with robust health care and well-being. This was in essence possible because of exploiting molecular bonding among different metals, alloys, and other chemical compounds. Studies concerning metals along with the understanding of their ductility, malleability, and thermal conductivity aspects lead to the adoption of different metals for different purposes and to the development of the discipline of metallurgy. This formed a basis for industrial applications and the development of gadgets for day-to-day living. Deeper scientific studies of the associated features of thermal, electrical, and electronic properties revolutionized the growth of technological innovations in the medical, communication, and transport industries in the last one hundred years. More advanced studies in these disciplines have led to innovations in micro and nanotechnologies which have opened up immense possibilities of applications in almost every walk of our lives.

Recent developments in carbon nanotubes have helped create ultra-strong and ultra-light-weight materials that find use both in the aircraft industry and Space Technology.

The origins of developments in the science of genetics are attributed to Gregor Mendel (1866) whose experiments on various types of peas led him to speculate on laws that could explain the transfer of characteristics from parents to off-springs. His observations formed a basis for the science of heredity and genetics, developed later by Hugo de Vries, X. Y. Correns, and E. T. von-Sysenegg in the 1900s. Erwin Chargaff discovered the equal ratio pairing of nucleotides and the X-ray diffraction studies of Rosalind Franklin and colleagues (1951) provided the helical structure of the DNA as propounded by Crick and Watson (1953), using Chargaff's data (Pray, 2008). Rosalind Franklin gave the understanding of DNA replication and hereditary control of cellular activities that led to the development of genetic engineering and technology. Amongst pioneering and significant discoveries that followed the DNA structure was that of PCR (polymerase chain reaction) (1983) in DNA technology. This opened up the technique of DNA sequencing and along with recombinant DNA technology this gave impetus to the field of genetic engineering which has a great impact on agricultural, medical, therapeutic, forensic, and genetic profiling (Winchester, 2020). Gene expression is a unique way of characterizing the adaptation of cells. Measurements of their levels, when exposed to a chemical, can be used as a 'genetic signature' for the identification of toxic products. This has led to genetic technology being used as a standard tool in molecular toxicology. It is indicated that the development of stressor-specific signatures in gene expression profiling will have a major impact on ecotoxicology. This development could have applications in pollution control and other environmental studies.

Louis Pasteur developed the method known today as Pasteurization, which is a must for our intake of milk as it destroys the disease-causing micro-organisms and also permits storage for a longer duration. Not many may know that the process he invented was to save the wine industry in France in the 1850s, and the problem was due to the high degree of fermentation caused by micro-organisms (microbes). His studies were aimed at understanding the structure of crystals which had features of both left and right facets that rotated the polarized light in both directions and these studies lead to the development of the discipline of stereochemistry (Philip, 1961).

Chronologically speaking, one can see that the scientific revelations that occurred in the 18th and 19th centuries gave the basis and impetus for a large number of discoveries and inventions that aided the development of modern science and technology. One can consider the various important aspects made by the discovery of the theory of electromagnetism by James Clark Maxwell (1865) which eventually led to the understanding of light as an electromagnetic wave. The understanding of the spectral features of elements and compounds that led to the understanding of the physical and chemical properties of matter, and the classification of plant and animal cells and tissues. This was possible due to the already available 'Microscope' which led Darwin to his theory on the origin of species. Apart from the basic understanding of most of the scientific theories, this period also showed the interrelationship among the physical, chemical, and biological features of matter which helped the understanding of the impact of geological and environmental conditions on life in all forms.

Serendipity in Science

Unintended discoveries are called 'serendipitous', where one finds something completely unexpected and rich without particularly looking for it. (The term is supposed to be coined from the word 'Serendeep', the name given by the British, to what is today 'Sri Lanka', who discovered it without looking for it and whose original name was 'Swarnadweep' meaning island of gold. There seems to be another explanation regarding the etymology of the term 'serendipity' which was supposed to have been coined by Horace Walpole in 1754 for an ability possessed by the heroes of a fairy tale called The Three Princes of Serendip who were making unexpected discoveries by accidents and sagacity (Wikipedia S).

The discovery of the antibiotic Penicillin by A.Fleming revolutionized the entire medical industry in the days to come and today antibiotics have become a household product. Alexander Fleming (1881–1955) was concerned about the death of soldiers during World War-I, more because of sepsis resulting from the infected wounds than the killings in the war. He realized that as the antiseptics then known worked only on the surface of wounds and did not penetrate to neutralize the microbes, he worked on finding the antiseptic that could kill the pathogens. In 1928 he studied the variation of Staphylococcus aureus which is a bacterium that grows under natural conditions. Fleming had inoculated staphylococci on culture plates and left them on a bench before leaving for his holiday. On his return he noticed that one culture was contaminated with a fungus, and the colonies of staphylococci immediately surrounding the fungus had been destroyed, whereas ones farther away were unaffected.

This gave birth to the discovery of Penicillin, the antibiotic he was searching for and he seemed to have expressed his delight with the words,

> One sometimes finds, what one is not looking for, as on the morning of September 28, 1928, I certainly did not plan to revolutionize all medicine by discovering the world's first antibiotic or bacteria killer, but that was exactly what I did.

However, it is also mentioned that (just to keep the facts clear), Fleming was not able to extract any usable amount of penicillin and it was H. Florey an Austrian pharmacologist and colleagues in 1939 who were successful in figuring out a way of purifying it for the use by the masses (Lorch, 2015). It may be mentioned that before Fleming, Edward Jenner (discoverer of smallpox vaccine in 1796) and Humprey Davy (who identified the 'laughing gas as an anaesthetic in 1799) are also cited as being serendipitous in their discoveries. The impact of these scientific breakthroughs in the pharmaceutical and medical industries is very well known. Among the more recent serendipitous discoveries, conductive polymers, which are organic molecules that can have metallic conductivity or behave as semiconductors have found immense application in the electronic industry including in the discovery of 'organic light-emitting diode' (OLED) used now in every household.

Until 1987, billions of batteries that had been marketed in myriad sizes and shapes depended upon chemical reactions involving metal components of the battery. But today a revolutionary type of battery is commercially available which stores electricity in plastic. The development of plastic batteries was also serendipitous and began with an accident. In the early 70s, a graduate student in Japan was trying to repeat the synthesis of "polyacetylene", a dark powder made by linking together the molecules of ordinary acetylene welding gas. After the reaction was over, instead of a black powder, the student found a film coating inside of his glass reaction vessel that looked much like aluminium foil. He later realized that he had inadvertently added more than the recommended amount of the catalyst to cause the acetylene molecules to link together. News about the foil-like film reached Alan Mac Diarmid of the University of Pennsylvania who was interested in non-metallic electrical conductors. Since polyacetylene in its new guise looked so much like metal, MacDiarmid speculated (1977) that it might be able to conduct electricity like metal as well and the investigators confirmed that polyacetylene exhibited surprisingly high electrical conductivity, and this accidental discovery brought in the technology of LED which have now been mainly used for innovations in display monitors, lighting and several other areas. Over the last few years, this technology has been the basis for thinner and thinner mobile phones. It is indeed becoming a mainstream display technology as OLED enables display panels that offer the best image quality and free design as they are flexible, transparent, efficient, and require a relatively small amount of power for the amount of light they produce.[3]

Literature is replete with numerous discoveries in science which all led to the use of technology that pervades our life. We rarely realize the kind of scientific rigor that underlies the development of numerous innovations which one takes for granted. The industrial world came into being and progressed only on the ingenious use of fundamental science and its applicability. We will be referring to various scientific discoveries and their applications in technological advancements as we proceed looking at the different segments of learning and practicing science. Before discussing the methodology, language and associated aspects let us take a look at the possible day-to-day events and scientific principles associated with them.

References

BSS. (2017). *A brief history of science*. Breakthrough Science Society.
Genta, G., & Riberi, P. (2019). *Technology and the growth of civilisation*. Springer.
Kak, S. (2005). *Science in ancient India*. https://www.ece.lsu.edu/kak/a3.pdf

---

[3]For details one can see: Wikipedia contributors. "Light-emitting diode." *Wikipedia, The Free Encyclopedia*. Wikipedia, The Free Encyclopedia, 22 Feb. 2022. Web. 4 Mar. 2022.

Knoblauch, J. A. (2020). *Environmental toll of plastics*. https://www.ehn.org/plastic-environmen
    tal-impact-2501923191.html

Lorch, M. (2015). https://theconversation.com/five-chemistry-inventions-that-enabled-the-modern-
    world-42452

Philip, C. (1961). *Giants of science*. Pyramid books, Grosset and Dunlap Inc.

Pray, L. (2008). Discovery of DNA structure and function: Watson and Crick. *Nature Education,
    1*(1), 100.

Shaver, P. (2018). *The rise of science*. Springer.

UKSC09. www.sciencecouncil.org/about-science/our-definition-of-science

Wikipedia contributors. *"OLED,"* *Wikipedia*. The Free Encyclopedia. Accessed October 19, 2021.,
    from https://en.wikipedia.org/w/index.php?title=OLED&oldid=1050635815

Wikipedia contributors. (2021). "Liquid crystal." *Wikipedia, The Free Encyclopedia*. Wikipedia,
    The Free Encyclopedia, 17 Oct. 2021. Web. 3 Dec. 2021.

Winchester, A. M. (2020). *Genetics*. 2020, https://www.britannica.com/science/ / genetics

Chapter 2
Science in Daily Life: Can One See Science in Routine Actions?

Introduction

With the advent of new technologies and several methods of analyzing the events that happen around us, one can identify a few scientific concepts that one encounters in daily life. The basic approach to science is experimenting, questioning, reasoning, and arriving at logical conclusions, it is useful to look for a scientific explanation to many of the events that occur in day-to-day living. Very familiar and commonly used terms like, inertia, gravity, work, energy, force are all concepts that are easily grasped and explained through very frequent happenings in our daily routines. Using proper scientific reasoning for some of the daily actions will help in understanding the how and why of events that follow one another. Once observed and analyzed, these concepts can lead one to look for events beyond and may help to follow the cosmic order to understand why it is so.

As often mentioned, science is a process of understanding through analysis of phenomena that happen around us in time and space. With this proviso, one should expect to explain the events of daily life where one is interacting with his/her surroundings. It should be a matter of routine to look at some of the day-to-day happenings and relate them to concepts in science. Being in the twenty-first century one must realize how fortunate are we to be in a world full of gadgets that make our lives far more comfortable as compared to that of our ancestors. Perhaps this extra comfort has lead us all to sluggishness? It is worth also thinking about the natural resources that are being consumed while creating extra comforts for ourselves. Starting from the mobile phones, two and four-wheelers, TV s and DVD s life, especially for those in urban areas with varied facilities has been very comfortable. While using all these technologies which are now being taken as necessities it should be obligatory for one to think about the efforts that went into creating these gadgets and the scientific principles that initiated and fostered such technologies without forgetting the resources that were used while creating.

© The Author(s), under exclusive license to Springer Nature Switzerland AG 2022
A. R. Prasanna, *How to Learn and Practice Science*,
https://doi.org/10.1007/978-3-031-14514-8_2

Before going into the physics or engineering required for such technologies let us consider simple everyday facts that inform on several scientific principles and explanations that govern the phenomena which are often taken for granted. Doing science need not always be probing deep philosophical issues or hard-core technological innovations but could simply be finding out why or how several of our common experiences occur.

As already hinted the basic methodology for science comprises experimenting, questioning, reasoning, and then arriving at logical conclusions that could lead to further questions. This process continues and leads one to a deeper understanding of the phenomena and the principles that govern the processes around us. Scientific reasoning leads to and is led by a continued process of thinking. As aptly said by Einstein, "the whole of science is nothing more than a refinement of everyday thinking. It is for this reason that the critical thinking of the scientist cannot possibly be restricted to the examination of the concepts of his own specific field. He cannot proceed without considering critically much more difficult problem, the problem of analysing the nature of everyday thinking."

It would be appropriate to extend the last sentence as 'nature of everyday living and thinking'. From the time we get up in the morning many of our actions if examined logically comprise a good deal of scientific reasoning. Most of the time one works through reflex actions but a retrospective analysis of 'why did I do so', would lead to an explanation that could be scientific. At times one follows instructions by elders cautioning of the consequence of some actions which could be coming as a result of their experiences. Still, situations could arise where a more critical mind not willing to accept the conventional wisdom may lead to the development of alternative hypotheses or explanations, the success of which can lead to a new understanding. This requires one to have an open mind while going through daily actions and be aware of any unexpected happenings that may not be common experience. Let us consider a few of the basic concepts encountered in daily life quite routinely.

Motion, Inertia, and Gravity

While traveling in a moving vehicle one experiences a jolt when the vehicle stops suddenly. If noticed carefully, the jolt is such that the upper portion of the body falls forward while the base remains static inside the vehicle. Similarly, when the vehicle moves suddenly from a stationary position, the upper portion of the body falls backwards. Both these occur because of *inertia*, which is defined as a state wherein a body continues in its state of rest or state of motion'.

How are these experiences explained by inertia? While traveling in a vehicle either sitting or standing, the lower portion of the body or the feet is fixed to the vehicle, whereas the upper portion is free but being dragged by the lower portion. Because of this influence, the whole body has a velocity in the direction of motion. Now when the vehicle suddenly stops, the lower portion in contact with the vehicle comes to a halt whereas the upper portion is still in motion due to inertia. That is why one falls forwards. Similarly, when the vehicle starts suddenly, the lower portion gains motion whereas the upper portion wants to continue in the state of rest. Consequently, one falls backwards. In motor accidents, when the drivers do not use seat belts, because of the crash the body surges forwards and the head or chest smashes against the steering wheel, causing serious injury. Seat belts keep the entire body fixed to the vehicle and thereby restrain the upper portion of the body from falling forward and avoid mishaps.

The word inertia comes from the Latin language meaning 'inactivity'. The word inert is so commonly used in English to indicate that someone is lazy by saying he/she is inert. One might have heard of 'inert' gases in chemistry like argon, krypton, xenon, etc. which are non-interactive with other elements and thus termed inactive or inert.

'Every body continues to be in its state of rest or of uniform motion unless acted upon by an external agency'.

Isn't this very familiar? Supposing one did not want to do any work and felt like dozing off when either the parents or the teacher (external agency) forced him/her to change from that position. When an external force is required to change the current state of rest and one's unwillingness to get active by oneself is due to inertia. What is this external agency? And what is its role? Can there be an example wherein the influence of the external agency is quantified? Consider a bicycle and a car. It is known that it is far easier to push a bicycle but very difficult to do the same of a car. When both are at rest, to make them move one has to overcome their inertia. It is also known that a single person can push a bicycle while moving a car needs many more. This is because a car is much heavier (more massive), and offers greater resistance than the bicycle which is lighter (less massive). This is analogous to pushing a small stone vs. pushing a larger rock.

These facts provide simple illustrations of the relationship between inertia and the mass of a body. Inertia is directly proportional to the mass of a body. The bigger the force, the greater is the acceleration imparted on a given body. The same force applied on two bodies of different masses, result in different effects, while one needs different forces for bodies of the same mass at different distances to have the same effect. One can easily experience this feature while playing the game of carom, where one needs strokes with different pressures for pouching pawns at different distances.

Concepts of inertia, mass, state of rest or motion, acceleration, and force are simple and occur so commonplace that they are almost taken for granted. Almost four hundred years ago it took the genius of Sir Isaac Newton to identify these properties of physical systems and to enunciate the three fundamental laws of motion viz,

1. Bodies continue to be in their state of rest or of uniform motion unless acted upon by an external agency. This is known also as *Galileo's law of Inertia*.

2. The external agency is called *Force* (F), which is proportional to the mass of the body (m) and induces acceleration *(a)*, to the body, i.e. *F = m a.*
3. Action and reaction are equal and opposite.

These three statements which almost are of daily experience are known as the laws of motion and they form the basis for Newtonian mechanics. This was the first fundamental theory in physics. Besides enunciating these laws Newton added another important concept which describes the major force that we all experience- Gravity defined as 'being proportional directly to the product of the masses of the two bodies and inversely to the square of the distance between them', familiarly known through the equation, $F = G\,m_1\,m_2\,/r^2$, G is called the Newton's constant of gravitation. It is one of the fundamental universal constants.

The force of gravity is very common and is experienced while climbing up a hill or a flight of stairs. It makes one feel tired due to the work one needs to do to counter the force of gravity. Similarly, when a ball is thrown up it reaches a certain height and then falls back to the ground. It is the same force that brings the ball down that makes one tired while climbing up. Though the effects of gravity must have been experienced by people for ages, it was not recognized till Newton in the year 1685 published his famous treatise '**Principia**' explaining what keeps the Moon in its orbit around the Earth or make things fall to the ground. The story of an apple falling is just incidental. Newton was wondering as to why the Moon is going around the Earth instead of crashing on it. He realized that the attractive force of earth's gravity has to be overcome by the moon's orbital motion with a certain velocity, which is the same reason for the planets going around the sun.

One sees pictures of astronauts flying about inside spacecraft holding the walls for support. Up there they cannot walk even if they want to, because the gravitational force does not exist inside the spacecraft when it is in faraway space. Gravity plays such an important role in our lives that without it one cannot think of doing anything, for example like pouring a glass of water from a jug or adding a spoon of sugar to a cup of coffee/tea.

Gravity is always directed towards the geometrical center of the mass creating it like the Earth on which we stand, sit or sleep. For every object living or non-living, there exists a center of gravity and as long as this center stays within the space covered by the body, the body stays in equilibrium. For example, when in a standing position, the perpendicular from the center of gravity of the body passes through the center between its legs and thus is in a stable equilibrium state. For any individual, while sitting on a chair with a straight backrest in an erect posture, it will be almost impossible to get up without either bending forward or pushing the feet below the chair. This is so because, in that sitting position, the center of gravity is located near the spinal cord at the upper back level. A perpendicular line from that point will touch the ground below the chair and away from the feet. Thus unless one shifts the position of the center of gravity, through bending forward or pushing the feet below the chair, one cannot stand up. It is often said that carrying a water pitcher on the head is simpler and more elegant for the gait, than carrying it on the waist with a hand surrounding it. This is again a judicious distribution of the load that keeps the

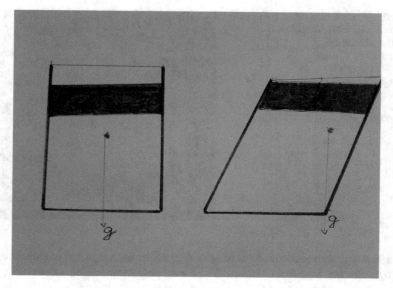

Fig. 2.1 water tank on the roof and the effect of earthquake

center of gravity appropriately within the limits of stability. Similarly while planning a building the structural engineer should work out the location of the center of gravity while planning the load distribution of the building. One may remember the tragic collapse of several high-raised buildings in Ahmedabad in the devastating Kutch earthquake of 2001. The earthquake resulted in great loss of life and property. It might come as a surprise to many, that the cause in some cases, was large water storage tanks on the roofs of some of those buildings, which caused a different distribution of weight and thus loss of equilibrium when the quake happened. In normal times, the design of a building is such that the line that passes through its center of gravity goes through the base of the building keeping the structure stable. When the Earth shook and the building wobbled, the building structure which is fixed to the ground moved with the base, whereas the water inside the tank which was free to move rushed completely to one side of the tank (Fig. 2.1) thus causing an imbalance in the load that led to the change in the position of the centre of gravity and consequently the building collapsed.

Gravity also determines the shapes and sizes of bodies along with their internal forces. Life on earth is completely dependent upon the value of 'g' the acceleration due to gravity on earth. The value of g on the surface of any body is $g = GM/r^2$, M is the mass, r the radius of the body, and G Newton's constant. (On the surface of the Earth, g = 9.8 ms/s^2). What changes would occur if this value was different is a subject of research and space missions are helping us to understand the effects of microgravity (value less than g) at regions far into space. Similarly, studies are made on the effects on astronauts of hyper-gravity (value > g) during the take-off of a rocket carrying the satellite (which could be almost 3 g). Similarly during re-entry as

Fig. 2.2 'Core element of
an electromagnetic levitator'
(photo credit: DLR).
V. Plester and T Russomano
(2020) 'Research in
Microgravity and Life
Sciences-an introduction to
means and Methods', DOI
10.5772/interchopen 93463

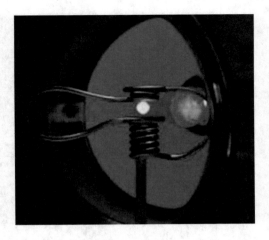

well astronauts experience effects of gravity ranging from 0 to 1.6 g. Effects due to microgravity are significant and could help in preparing certain alloys which are otherwise difficult to form on earth due to the normal gravity. Microgravity research is being carried out in areas of fundamental physics, material science as well as human physiology. An interesting and useful effect in microgravity is that the internal forces of a system like the elastic, cohesive and viscous forces may overcome gravity and allow new situations that help under different conditions. A practical utility could be for the liquid phase of materials which can be held in a crucible by electromagnetic or acoustic forces without the liquid touching the sides of the container as shown in the core element of an electromagnetic levitator (Fig. 2.2).

Microgravity research has opened up new possibilities for investigations in physical and life sciences. Such investigations are necessary to increase our knowledge of how physical systems and the human body functions under the change of gravitational acceleration on different space platforms which one may have to experience in deep space explorations. The differences can occur at all levels from the cells to whole-body systems in their reactions to adjust, perform, react, and function under microgravity. While experiments can be more easily planned through the use of a centrifuge to study the behavior in hyper-gravity it is difficult to set up the situation of being in microgravity on earth and thus situations are created in what are called parabolic flights where the aircraft is maneuvered from 1.8 g to zero-g and back to 1.8 g all lasting about 70 s (Plester & Russomano, 2020).

One of the most unexpected and useful results concerning gravity (because of the theory of relativity) that manifested in the mid-90s was the concept of GPS the global positioning system. Theory of relativity which was indeed considered as the purest of the theoretical developments displayed its full applicability in the technology associated with GPS. This is a satellite-based system that helps one to find her/his position anywhere on earth, depending upon the allowed frequency, to the accuracy of the order 10 ms, to less than a meter. The system consists of 24 satellites orbiting around the Earth, four in each of six different orbital planes inclined to the Earth's equatorial plane at an angle of 55°, carrying very highly accurate ceasium clocks, synchronised with one another. The positioning of the satellites is such that, from any point on earth, four satellites are above the horizon at all times. The satellites orbit at a height of 20,000 km above the Earth's surface, with a period of 11 h and 58 min (half a sidereal day) such that any fixed observer on the ground will see a given satellite, almost exactly at the same place on the celestial sphere, two times in a day. The satellites transmit synchronous timing signals that carry coded information about the transmission time and position of the satellite. The transmitted data are continuously monitored by receiving stations around the globe and forwarded to a master control station where the orbits and clock performance are computed and uploaded back to the satellites for retransmission to the users (Ashby, 2003). Apart from relativity, the science of atomic clocks and the knowledge about the frequencies emitted on change of energy levels (work of Neils Bohr and colleagues during early twentieth century using quantum theory) all come into play in the development of this technology. Today most passenger cars are equipped with this system. Thanks to GPS the international network of clocks is synchronized to an accuracy of a billionth of a second per day. One may wonder as to the role of the theory of relativity which happens thus: As time is not an absolute quantity, the time shown on a clock in a satellite undergoes two changes *viz.*, goes slow with respect to the observer on earth due to special relativity (by 7 microsecs) and time dilation and also goes faster (by 43 microsecs) in space due to the change in the gravitational potential (from ground to sky) resulting in a overall excess of 38 microsecs which is corrected constantly by the atomic clocks through on-board computers. Now a days GPS technology is being used by everyone working in different sectors from agriculture to aviation, defence to navigation and map-making. Geologists have used the system to monitor the continental plate movement which can give helpful information on possible earthquakes or tsunamis of magnitude 7 and above through the pattern search of the old seismic recordings (GPS 19).[1]

Work and Energy

On any given day there must be times when one feels tired and would not want to do any work. When asked to do some work, one might say, 'Oh no, I do not have the energy to do that. The two words work and energy are often used together. By definition, *energy is the capacity to do work* and it appears in several forms such as mechanical, heat, light, sound, and electrical. It is to be appreciated that life would not be possible without any one of these sources of energy being available. For every

---

[1] For details Wikipedia contributors. "Global Positioning System." *Wikipedia, The Free Encyclopedia*. Wikipedia, The Free Encyclopedia, 27 Feb. 2022. Web. 4 Mar. 2022.

Fig. 2.3 Setting on of Breeze. Credit: NOAA/"The Comet program" Article by: Nick Sharr (https://weatherworksinc.com/news/Sea-Breeze) Picture credit NOAA

action, energy is needed in one form or the other. A day begins with the morning tea/coffee which needs boiled water. One keeps the kettle on the stove and switches it on and the gas flame underneath burns providing a clear example of one form of energy converting itself to another form. The chemical energy of the gas gets converted to heat energy which accumulates at the bottom of the kettle and passes it on to the water inside. The water inside the kettle, comprising a large number of molecules gathers the heat energy which makes the water molecules move faster (as they acquire kinetic energy). As the water gets hot inside the kettle, water starts rising upwards. This is so because cold water has a certain distribution of energy of molecules and as it gets hot the molecules acquire more energy and start moving apart and the density at the bottom decreases. The cooler water on the top is still of higher density and therefore sinks downwards and the heated water at the bottom moves upwards. This process is called convection, the process of heat transfer in fluids. Often in the bathroom, one gets tricked by dipping a finger in a bucket of hot water to find it hot enough, but as one reaches the bottom of the bucket the water may be only lukewarm, as heavier density cooler water stays down. During morning walks before sunrise one normally finds the breeze to be absent but about half an hour after the sunrise a slow cool breeze sets on. The reason is as follows; the absence of sunlight during the night, cools the atmosphere and the air temperature is nearly the same from the ground level to the top. With sunrise, the Sun's radiation enters through the atmosphere and the ground starts warming up, and through conduction, the air at the ground level also warms up. This warmer air rises creating space below to which the denser cool air from a higher level in the atmosphere or from the sea surface descends and sets the convection of air. This results in the initiation of a cool breeze (Fig. 2.3).

On a global and monthly scale, these phenomena during summer set the weather pattern and bring the rain! The formation of rain consists of three stages,

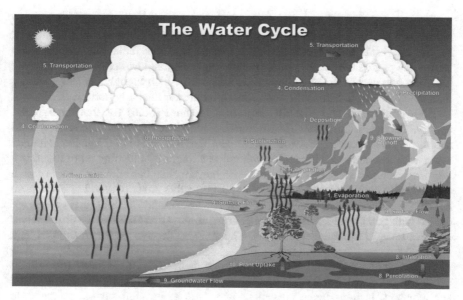

Fig. 2.4 The Water Cycle. Credit: Earth's water continuously moves through the atmosphere, into and out of the oceans, over the land surface, and underground. (Image courtesy NOAA National Weather Service Jetstream.) https://earthobservatory.nasa.gov/features/Water/page2.php

evaporation, condensation, and precipitation which are supposed to be part of the water cycle. The summer sun heats the land and makes the air density thin above the ground and also it evaporates the water in the oceans. The hotter air and the water vapor both rise upwards. When the vapor rises above the dry air, the surrounding air at higher levels of the atmosphere which is cooler cools the vapor. The reduction of temperature consequently condenses the vapor back to liquid and it forms small droplets which due to their cohesive nature combine to form the raindrops. This is the process of precipitation. This cold moisture-laden air from above sea level moves towards the land bringing in the clouds and rain. The process also requires the sea surface temperature to be within certain limits for the efficient formation of clouds (Fig. 2.4).

One might have often seen the pictures of cloud cover taken by the remote sensing satellites, generally shown in weather forecast news. One must have also heard of the term 'acid rain'. What is that? We all know how pollution is affecting modern life because of the excessive release of industrial waste to the atmosphere and rivers as well as the excessive burning of fossil fuels like oil and natural gas. As discussed at various levels (both scientific and political) while on the one hand, it increases the carbon and nitrogen in the atmosphere, the industrial waste gases add sulfur dioxide and nitrogen oxides to the atmosphere. When these oxides mix with water vapor they form sulphuric and nitric acids which are very harmful. When it happens during precipitation, the raindrops carry these acids and form acid rain.

The process of land warming from sunlight is through the transfer of heat by radiation. One must have heard of people using radiation therapy for backache or knee aches through the use of an infrared lamp. The heat radiation from the lamp warms the part of the body to make the muscles relax and be comfortable. Infrared radiation is a part of the electromagnetic spectrum which though has lesser energy than the visible light carries the heat from the source to the body portion effectively.

Both convection and radiation require a medium (fluid) for transferring heat energy. There is a third way of heat transfer—conduction, which occurs in a solid medium by direct contact between molecules. This method of heat transfer is indeed faster than the other two modes. Burning one's finger by touching a hot metal is a very common experience. It is to avoid such burning, handles of many vessels which are generally kept directly on stoves, have a protective cover of wood or ebonite on its handle, which are poor conductors of heat. The bottoms of some such vessels are coated with good conductors like copper or aluminium that facilitate the efficient and uniform transfer of heat to the vessel. The story so far was on the transfer of heat energy from one place to another.

Another common form of energy that one deals with in everyday life is light energy without which it is very difficult to work at night. During the day we have sunlight and its importance is well known. Due to its high surface temperature (~5800°K), Sun emits radiation in all frequencies, from long wavelength radio waves to X-rays, but peaking in the ultraviolet. Of these only the infrared, the visible (and a small portion of UV) reaches the surface of the Earth, rest being absorbed or scattered by the atmosphere. UV is absorbed by the Ozone layer. The white light from the Sun is a mixture of colors, as pointed out by Newton, which are radiation at different frequencies. Interestingly enough the way these radiations carry energy is in small packets called 'quanta' as pointed out by Max Planck (1900) or photon as identified by Einstein, such that each photon carries its characteristic energy. Though initially there was a difference in the understanding of the nature of light (whether it is a wave as suggested by Huygens or a set of particles as told by Newton) after the discovery and theoretical explanation of the photoelectric effect one understood the dual nature of light. This effect is responsible for your viewing of pictures in a TV. Photons from pictures transform their energy to electrons in some metals and these electrons get free and as an electromagnetic wave travel from far away studio carrying the digital information. These waves when received by the antenna of the TV in your drawing room get converted back into pictures (through the electron tube) on your TV screen. The new technologies in this regard have given modern television with LEDs and plasma screens of much higher quality pictures. The photoelectric effect was explained by Einstein in 1905, for which he got the Nobel prize in 1921. The effect itself was discovered by H. Hertz in 1887. Hertz also was the first to produce electromagnetic waves, which was identified through Maxwell's theory of electromagnetism that forms the basis for the entire communication technology.

As mentioned above, the sunlight is composed of electromagnetic radiation of different frequencies (or energies) spread over a large spectrum ranging from radio waves to X-rays, with visible light (to which the human eye is sensitive) being the

THE ELECTROMAGNETIC SPECTRUM

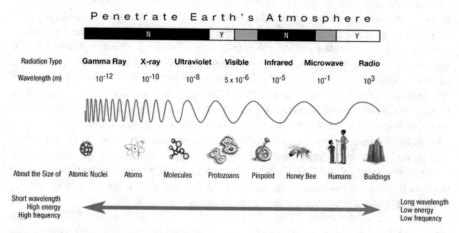

Fig. 2.5 Credit: My NASA DATA Sponsored by NASA. https://mynasadata.larc.nasa.gov/basic-page/electromagnetic-spectrum-diagram

VIBGYOR spectrum consisting of the seven colors viz. violet, indigo, blue, green, yellow, orange, red in specific proportions classified as the optical window. Waves with wavelengths lower than violet are the ultraviolet (UV), X-rays, and Gamma rays, whereas those with wavelengths greater than red are Infrared (IR), Microwaves, and Radio waves. Today one has astronomies associated with each one of these frequencies of electromagnetic waves that have provided the view of the Universe in the entire electromagnetic spectrum (Fig. 2.5).

As is known, human eyes are sensitive only to the visible part of the electromagnetic spectrum whereas the eyes of insects and many other animals are sensitive to infrared too. Thus they see clearly in the night when the normal light is not there. The use of infrared cameras for space missions and night warfare is well known. One might worry whether the reception of high-frequency radiation may cause harm to living beings. Luckily nature has been so kind that it has protected the Earth with a shield of the Ozone layer which absorbs the UV radiation and a magnetic field to deflect away the high-energy particulate matter from reaching the surface of the Earth. That is why astronomers need the aid of high-flying balloons going beyond the atmosphere or the rockets and satellites to receive cosmic radiation of higher energies than the Ultra Violet.

The use of bulbs and tube lights for illumination at night is a common practice. Both deliver white or near-white light but work differently. In an ordinary electric bulb (generally known as an incandescent bulb) there is a filament (generally of tungsten because of its very high melting point) which when supplied with electric current, gets heated and glow producing both heat and light energy. A significant amount of energy is lost as heat as this is not used for any helpful purpose. The use of a tube light minimizes such wastage of energy as it works on the principle of fluorescence and plasma emission. Tube light is an electron tube (with a little argon gas and mercury vapor) that has also a filament that produces streams of electrons when current is passed through it. These electrons ionize the argon atoms in the tube which in turn ionize the mercury vapor in the tube which becomes a plasma. Plasma is indeed a state of matter (called the fourth state of matter) which is a collection of ions and electrons. This plasma emits ultraviolet radiation which in turn excites the fluorescent material coated inside the tube which then emits the visible light. Though this process takes a few microseconds for the light to come on as compared to an incandescent bulb, there is no energy loss in the form of heat. As filaments in bulbs burn out faster, the life of an electric bulb is much shorter as compared to that of a tube light. Nowadays one has LED (light-emitting device) lamps which are long-lasting and more economical as they consume much less power as compared to normal bulbs and tube lights. Further, the reduction in power consumption is very important from the environmental point of view as the production of electric power drains our fossil resources that can harm future generations.

As is known the food one eats come from plants. The preparation of food by the plant is called 'photosynthesis', which is synthesizing (preparing) food by photons. The leaves of the plant use the constituents, water from the roots and carbon dioxide from the atmosphere, and the green pigment 'chlorophyll' absorbs photons from the sunlight and the food gets prepared in the leaves (kitchen of the plant).

A fundamental law of nature is that 'energy can neither be created nor destroyed' (law of energy conservation). It can only be converted from one form to the other. In any substance or body, energy is stored as potential energy which gets converted to kinetic energy when the body is in motion. Thus the total energy of a body is the total of its potential and kinetic energy. While performing any work, muscles are used as these have stored mechanical energy derived from the food which has chemical energy that gets converted and stored in the body as mechanical (potential) energy. Physical tiredness after some work arises from our spending more energy than the body could afford, and therefore the body needs to be supplied with more energy through food intake and rest. Walks imply work against gravity and this requires converting our potential energy to kinetic energy. Therefore, when a fat person and a lean person walk the same distance or do the same work, the fat one would get tired early as he/she has to spend more of the potential energy because of the extra weight. The simplest example for understanding the relationship between work and energy is by trying to lift a heavy stone from the ground. As the stone is linked to earth by gravity (holding the stone firmly) one has to do work to lift it against gravity and this requires energy. One knows that the difference between a strong person and a weak person doing the same work is in their use of different amounts of energy.

By definition, mechanical energy is the energy that is used to set things in motion, through the conversion of potential energy to kinetic energy. In riding a bicycle, one pushes the pedal up and down, or in rowing a boat one swings arms forwards and backward and converts the stored mechanical (potential) energy into kinetic energy, leading to the motion of either the bicycle or the boat. Oft used statement of wasting of energy, does not imply the destruction of energy but only means that the energy has been converted into a form that is not usable. For example, cooking on a gas stove or an electric stove converts the chemical energy of the gas to heat and light or converts the electrical energy to heat. If there is nothing on the stove, either the chemical energy of the burning gas or the electrical energy is lost through radiation which is a waste of energy. In the case of petrol-driven vehicles, the engine combusts petrol to convert the chemical energy to mechanical energy, which when transferred to the wheels sets the vehicle in motion. While at a stop at a red light, if the engine is not switched off the energy produced is not used for motion and hence leads to a waste of petrol. One of the common examples of wasting energy by most people is by not switching off the lights and fans in workplaces while leaving the rooms. Extreme care is needed as these conventional forms of energy are produced from precious and limited natural resources that will soon exhaust and will put future generations into difficulty.

Current talks about non-conventional energy sources relate to solar energy, wind energy, and tidal energy. Such natural resources arise from the Sun and will not exhaust as long as Sun is shining which will remain so for several hundreds of millions of years. Conversion of these forms of energy to electricity or heat is indeed the most useful aspect of research that could save humanity. However attempts to use these resources are still on a limited scale through solar -panels, -cookers, -heaters, windmills, and tidal machines. The entire space exploration depends on solar energy. The power supply to satellites when in space, is from solar panels on the satellites which convert the solar radiation (light) to electrical energy to power the instrumentation on board. It will be prudent for countries like India with almost 330 days in a year of solar radiation to develop varied means of using solar energy more effectively. As an aside, it may be pointed out that even the process of drying wet clothes in the Sun is a way to use solar energy. One other source of energy production that is being researched for the last eighty years or so is fusion energy which if realized effectively, will be a boon to humanity.

The kitchen at home is a good science laboratory as several of the procedures followed are similar to doing experiments. During cooking of vegetables in an open vessel, the salt is added only after cooking for almost half the required time. This is so because cooking vegetables is like the process of osmosis and thus only when the concentration of the cooking water is less than that of the juices inside the vegetable, water can enter into the vegetable and make it soft (cooked). What is osmosis? This is a natural process where two fluids of different concentrations when kept together separated by a semi-permeable membrane, the one with a lower concentration diffuses through the membrane towards the other with a higher concentration. A salt solution is normally of higher concentration than the juices inside the vegetables. Thus to ensure that the vegetable is cooked tastily and rendered soft, salt is added to the water outside later in the process. There is also another reason for the addition of salt in the cooking water at the appropriate time and that seems to be to keep the right amount of balance of salts inside as well as the outside, particularly for the type of vegetables which are rich in mineral salts. Hard grains take more time for cooking as compared to vegetables which have thinner skins. Even while mixing the constitu-ents, the measures are important because the uniformity in taste occurs only when the molecular mixing of the constituents follows a definite pattern. However as modern cooking uses mostly pressure cookers, wherein the steam pressure built up inside helps the water to get into the vegetables one needs to check out carefully the correct procedure to be adopted. Good cooks follow these points empirically and through experience, but the whole subject of food science scientifically deals with such issues, ensuring reproducibility and deliciousness of the final product. A healthy human body requires apart from carbohydrates and proteins variety of mineral salts such as iron, calcium, magnesium, phosphorus, sodium, and potassium for its chemical activities and construction of tissues. A healthy daily diet, therefore, is a must with all these in minute and proportionate quantities. Consumption of fruits and vegetables provides many minerals needed. Often those who sweat more need electrolyte-rich water to compensate for the loss of mineral salts through sweat and therefore are advised to take dosages of the electoral powder in water or fruit juice. It is a little amusing to find that nowadays commercially something called 'sports water' is sold and sports persons advertise this product which is just mineral water added with electoral salts.

The process of digestion also is a chemical process with the body fluids interacting with the food consumed throughout the alimentary canal, liver, and intestines, starting with saliva in the mouth. Understanding these features of the functions of the human body is advisable for maintaining a good and healthy life. One may say that in normal daily life checking out and being advised on how to eat, when to eat and what to eat also do have scientific reasons.

If only we learn to look around properly and carefully one can see the results of science and science-based technology in every sphere of our life. As mentioned, the discovery of the wheel gave impetus to the entire system of transportation both of humans themselves and the goods they needed for life. Starting with simple carts and animal-pulled chariots, one developed more easily manageable bicycles. Learning about steam pressure, James Watt invented the steam engine, and this eventually lead to the development of most of the modern transport facilities apart from initiating the industrial revolution. These developments went hand in hand with the discovery of diesel and petrol engines which are the basic elements of modern transport systems. With this advancement, the movement of people from one place to another became a simple reality which in turn brought in increased interactions among the people across cities and nations. These interactions have had a great impact on human society as travel has become an absolute necessity both for work and entertainment.

Along with transportation, the other important aspect of our lives is communication. The changes that have occurred in the last one hundred years in communication are phenomenal. Gone are the days when one waited for weeks to get a response for a letter posted in one part of the world to another. The telephones and the wireless had brought in closer interactions of people which in the last few decades has reached a monumental level with the arrival of computers, the internet, and mobile phones. Thanks to advances in space research and technology, communication satellites have revolutionized the daily life of a common man in every way possible. Apart from these, science has brought about miracles in the fields of medicine and surgery. Several discoveries and inventions in the areas of chemical and biological research have yielded medicines and methodologies for curing as well as preventing the spreading of diseases. It has also made it possible for the blinds to see, for the deaf to hear and for the crippled to walk apart from improving the general health and wellbeing of humanity and increasing longevity. Surgeons today can see the internal organs even before they make the first incision on the body, made possible by X-ray, MRI (Magnetic Resonance Imaging), CAT (Computerised Axial Tomography), Endoscopy, etc. It has also become possible in some cases to use laser beams (mostly eye surgery) instead of knives and thus avoid blood loss. Of all the medical facilities the organ transplantation, particularly for the Heart and Kidney is considered the greatest benefactors provided by science-based technology to human society.

Geophysical and geochemical research has yielded a better understanding of our earth and its surroundings which in turn is very helpful in looking for underground resources, controlling pollution, help maintain ecology and the environment. Feedbacks from satellite-based research on these topics have been very helpful in weather prediction and climate studies.

Looking back at the industrial revolution that changed the life style of humans it is quite revealing to notice a direct relation between deeper understandings of science leading to new technologies that had a bearing on industrial applications. The concept of alloys, through mixing metals to increase their ductility and malleability was one of the early scientific notions that helped in the middle ages to produce weapons as well as utility products is one of the finest examples. As pure iron was not useful for making tools, mixing it with carbon revealed the significance of steel which then dominated as the most used alloy for use in almost all industrial applications. As narrated (Genta & Riberi, 2019) in 1730, the first blast furnace was built in Britain which produced cast iron but it took almost a hundred years before the discovery of converters that made production of steel effective and as a by-product produced the 'lighting gases' which was used for house and street lighting. Further steel was used in the construction of machines which were lighter and better apart from using it in construction of bridges and buildings. Continuing, they point out that the new steel obtained through the Bessemer converter replaced the old 'puddled iron' of the previous century and further theoretical developments in chemistry and metallurgy revolutionised the resources for industrial applications. The production of newer alloys of aluminium, magnesium and titanium also joined the developments of synthetic materials such as thermoplastic and thermosetting resins and these together with steel have been in the service of humanity during the last couple of centuries.

Of the newer discoveries and applications in the last hundred years the most significant are the computers, the digital language and their many faced utilities. The two important names that need to be associated with this developments are of Charles Babbage (1833) and Alan Turing (1936). While Babbage developed the input and output devices for a computing machine in the form of data on punched cards, Turing introduced the concept of 'programming' the executable instructions that has today become universal. Turing's idea could find success mainly due to the associated development of the electronic digital machines which replaced the older mechanical analogue machines. Further this whole exercise found a very happy and useful structure in the idea of 'integrated circuit' in 1958 by J. Kelly (Nobel prize 2000) who invented the 'chip' that could carry the information (data) coded in the integrated circuit (Shaver, 2018). Rapid progress in the electronic industry over the last fifty years has helped particularly the computer industry which in this time has been able to reduce the sizes of computing machines. This has resulted in a big way for the air craft and space industry as they can be equipped with miniature computers that guide the system with on board computations of the varying data of the environmental conditions through their flights. The programming aspects of the mathematical computations required in solving complicated non-linear coupled equations (both algebraic and differential) that govern such situations has been made possible through important advances in the understanding of the theory of Chaos, analysis of initial and boundary value problems, logistic maps, neural networks and several others which in fact identify the role of mathematics in learning science and its applications. A very important feature that has made the use of computers as almost a daily necessity is due to the developments in computer graphics. Be it working from home (that became a necessity during the unfortunate pandemic affecting the world) or playing games, or communicating with friends face-to-face, computers are now unavoidable. Apart from this, computer graphics play a very important role in designing almost any product through computer-aided design (CAD) which are very useful in architecture and planning, product design, modelling exhibits, interior decoration, gaming industry augmented and virtual reality, and 3-D printing and a few others.

Continuing on one can explore the various day-to-day events (normal and unusual) where science is entrenched either in the cause or in the effect. Science simply explains reasons (how and why) for all our actions and reactions in daily life and technology based on scientific principles provides all the comforts we enjoy.

References

Ashby, N. (2003). Relativity in the global positioning system. *Living Revs in Relativity, 6*, 1 (online article) March 2016. http://www.livingreviews.org/lrr-2003
https://smd-prod.s3.amazonaws.com/science-red/s3fs-public/thumbnails/image/EMS-Introduction.jpeg
Genta, G., & Riberi, P. (2019). *Technology and the growth of civilisation*. Springer.
Plester, V., & Russomano, T. (2020). Research in microgravity and life sciences-an introduction to means and methods. https://doi.org/10.5772/interchopen93463
Wiki GPS19. https://bigthink.com/hard-science/predict-earthquakes-earlier/

Chapter 3
Scientific Methodology

Introduction

Learning and practicing science has a methodology. As it is a process of thinking about the knowledge accrued and wanting to be enhanced, one needs to look out for new features that might have appeared in one's experience which need to be analysed or reanalysed. The Greek school which is credited with developing a method for doing science dispensed with the idea of 'spirits' and started looking for logical reasons that brought in the cause-effect relation for explaining the occurred phenomena. Though the idea of causality was getting its foothold on scientific research the Aristotelian approach of learning from the writings of earlier masters was still very active. As explained by Immanuel Kant later in the eighteenth century the philosophy of getting new knowledge both through reasoning and through new observations gained momentum and established a definite methodology for doing science. Answering the questions What, How, and Why at appropriate levels depending upon the age and experience of learners sets the methods considered on a firm footing. It is suggested that learners should try and consider the life and work culture of the founding personalities involved with discoveries and inventions as a part of the training required to pursue scientific research. Scientific reasoning leads to and is led by a continued process of thinking which also includes a certain amount of intuition. Identification and a detailed study of any problem taken up for investigation followed by discussions with peers and experts form the foundations for new learning. Discussing the results obtained with others and verifying them thoroughly are very important features of scientific methodology. It is very reasonable to assume that already acquired knowledge initiates the thinking process.

In this chapter, an attempt is made to discuss how one gets at a deeper understanding of new observations and explanations of events that may arise out of different circumstances. Science is defined as gaining knowledge to understand the Universe and the principles that govern all its functions. Thus there ought to be a method for doing science and its methodology should be accessible to all and be repeatable. In the early history of science, the gathering of facts that had a bearing on natural phenomena was meticulous and tedious as these required a constant search and systematic observations or experimentation with very limited facilities. The reason for the methods of science being difficult is because any result obtained either by experiment or by observation has to be such that later investigators should be able to get to similar conclusions. This reproducibility is checked several times over by different individuals at different times and different places. Only then the results and their interpretation are acceptable as the possible fundamental truth. It is also said that science is an activity of creative and imaginative human beings, qualities which are themselves guided by discipline and self-criticism.

Early Beginnings

Historians of science (BSS, 2017) often credit three individuals as the pioneers that set a path for systematic science. They are

Galileo Galili (1564–1642), Francis Bacon (1561–1626) and Renè Descartes (1596–1650).

According to some historians of science, before these three pioneers experimental verification was not being considered as important, and only experience and the knowledge one had from earlier teachings were all that mattered. A systematic understanding of what was learned through readings from the records of the earlier generations was considered adequate by the philosophers of the early periods (circa 500 BC). This practice may be evidenced from the works of the Greek school and the example of the development of the geocentric universe theory. As the Greek philosophers began to rationalize all observations without recourse to spiritual influences, the motion of the objects in the sky like the sun, moon, planets, and of stars, posed a problem. Plato had tasked the students of his academy to develop a framework that would explain the motion of these celestial bodies. Before Plato and his disciples, the development of science had begun with the Greeks of the Macedonian peninsula who had elaborate mythology with spirits and gods, but were the first to get curiosity better of them to seek answers as to why things behaved as they did and this led them to make a transition from reliance on myth to the search for logical explanations—the knowledge! They sought to find general patterns in Nature and made an effort to determine their Order! Only around 600 BC, the first signs of systematizing the knowledge gained ground. This was with the arrival of Thales of Miletus (625–545 BC) and Pythagoras of Samos (570–495 BC) followed by the famous school of Plato (428–348 BC) and Aristotle (384–322 BC). Apparently, it was Thales who initiated the idea of the Universe working on natural rather than super-natural order and as such was considered the founding father of natural philosophy and seems to have had a profound influence on other thinkers of western philosophy. Pythagoras, well-known for the theorem on the property of a right-angled triangle, seems to have also contributed to the fields of music, astronomy, and medicine. He is attributed with the discovery of the relationship between the pitch and length of a vibrating string (learned through experiments) which made him speculate that all physical phenomena could be understood through fundamental mathematical relationships (Shaver, 2018). Eudoxus (400–350 BC), who belonged to Plato's school of thought, was a mathematician, who propounded the model of the geocentric Universe. This theory considers that the stars are hanging on the inside of a huge dark outer spherical shell which rotates once a day around the Earth east to west on an axis running north to south. Inside this sphere moved the planets, sun, and moon fixed in transparent shells rotating around different axes at different but constant speeds. In this approach by using 27 spheres, he explained the motions of the celestial system one for the fixed stars, three each for the Sun and the Moon, and four each for the five major planets, Venus, Mercury, Mars, Jupiter, and Saturn. Whenever new periodic phenomena were identified the system had to be expanded. Callipus, another disciple of Plato gave each celestial body an extra sphere, bringing the total up to 34, while Aristotle added a further 22 more spheres.

Aristotle (384–322 BC), considered by far the most prominent of all Greek philosophers was a man of vast curiosity with a wide-ranging intellect. He developed concepts on a grander scale than anyone before and was the first to conceive of an integrated system to explain how all aspects of the Universe worked together. Though it was later realized that many of his theories were incorrect or inadequate, he should be credited with the initiation of the development of a methodology for doing science. The Aristotelian view of the Universe finds a description in Dante's Alighieri, (Inferno) also known as the Divine comedy, with Hell at the center of the Earth surrounded by the planetary spheres and the Primum Mobile, the prime mover. Apart from the interest in developing Plato's school of thought of the Universe, he seems to have deeply studied many other aspects of physics, geology, zoology, medicine, and psychology. As will be discussed later this indicates the interdisciplinary approach to gaining knowledge. However, Aristotle's quest into the nature of Cosmos was mostly qualitative but not quantitative as it never used experimentation to confirm the ideas behind the thoughts and thus differed very much from the later viewpoint initiated by Galileo and followers. Some more disciples of Plato's school seem to have followed Aristotelian philosophy [Appolonius (~220 BC), Hipparchus (190–120 B.C.)], leading finally to Ptolemy (AD 100–170). Claudius Ptolemy, developed the idea of the geocentric theory of the Universe as was initiated by Plato's school, using the concept of epicycles basically to explain the irregular movements of planets which was more widely accepted. This idea dominated the thinking and beliefs for almost the next fourteen centuries, till the arrival of Nicholas Copernicus in the latter half of the fifteenth century, who developed the Helio-centric theory of the World expressing his view "*In the middle of all sits Sun enthroned. In this most beautiful temple could we place this luminary in any better position from which he can illuminate the whole at once? The Sun sits at the center, as upon a royal throne ruling his children—the planets which circle around him* (Fig. 3.1)".

Copernicus, after studying the various aspects of the then-popular Ptolemaic view of the Universe concentrated on making observations relating to the so attributed

Fig. 3.1 Ramesh Rao of Scion Advertising, for J N Planetarium, Bengaluru. Courtesy: J N Planetarium, Bengaluru, artist Ramesh Rao of Scion advertising

irregular motions of planets particularly Mercury, Venus, and Mars. As the idea of a heliocentric universe was in a sense due to Aristarchus almost 1500 years earlier, Copernicus tried to overcome the existing difficulty regarding the motion of planets with respect to the stellar background, by keeping the stars far away. Further unlike the older idea of having both sun and earth at the center Copernicus kept only the sun at the center and placed the planets including the Earth moving around the sun at different distances. Placing Mercury and Venus closer to Sun than the Earth and making the Earth move around, he could explain most of the difficulties associated with the geocentric theory. It is worth noting that despite working for almost 30 years, Copernicus was reluctant to publish what he had discovered as he thought that more observations were needed to be checked and rechecked before making it public. It is only because of George Rheticus, a German Professor who understood the significance of the work, Copernicus's heliocentric theory of the solar system was published in the year 1540. As one is discussing the scientific methodology this is a perfect example for scientific research which always has to be as accurate as possible. With meticulous observations of Tycho Brahe and the empirical laws of Kepler based on the Copernican idea which came a few decades later, the heliocentric theory got fully established.

It is only appropriate to mention that the developments in astronomy in ancient India were based solely on naked-eye observations of planetary motion aided by theoretical calculations. It is said that according to 'Vedic astronomy' practitioners counted months based on the phases of the moon and the motion of the Sun. It is interesting to find that as 12 lunar months (a lunar month is about 29.5 days, between two new moons) cannot give a Sun-year which is 365.25 days, the extra days were taken as the period when no auspicious event could be held. This practice continues even today. To make the two calendars synchronize, the Hindu calendar adds 1 month extra every 2–3 years and calls it 'Adhikmas'. Vedic mathematicians also identified the summer and winter solstices by the movement of the Sun towards the north and the south. They specified the positions of sun and moon in the background of constellations identified by 27 stars which were given specific names (BSS, 2017). Those interested in learning more on this subject could refer to (Kochhar & Narlikar, 1995; Iyengar, 2016).

Taking a look at the developments in a few other areas of knowledge during the early Greek period, one needs to mention the names of Hippocrates (460–370 BC) and Galen (120–210 AD) pioneers in the field of medicine. Hippocrates (known for the 'Hippocrates oath', which medical practitioners take even today) seems to be the leading figure who initiated the idea that diseases came from natural causes which require physical remedies and Galen discussing the human anatomy seems to have identified the fact that both arteries and veins carry blood. Among others, names of the early Greek philosophers Euclid (~300 BC) well recognized for his book on geometry—'Euclid's elements' and Archimedes (~287–212 BC) known for the principle he enunciated regarding the 'displacement of liquid by an immersing solid' are important in the development of scientific methodology. As one is not trying to narrate all the historical developments in science, there may be many facts not recalled here and are worthy of note, which one may look into in the earlier mentioned reference (Shaver, 2018).

Looking at the history of science from ancient Greeks up to Galileo and the modern era, one may see two distinctive methodologies that were followed.

1. The *relationship point of view*, like the one held by Plato and most of his followers, which holds the opinion that knowledge of Nature does not require observation and is attainable through reason alone.
2. The *empirical point of view*, as practiced by Galileo and his followers particularly Newton, which asserts the need to experimentation and observation.

Immanuel Kant (1724–1804) expressed that a successful methodology combines both the rationalist and the empiricist ways of thinking to arrive at a synthesis or a new point of view towards the nature of human knowledge. In this approach, there are both internal and external components to the way we understand Nature ahead of experience and through science. These are called *apriori*, or the pre-knowledge, and *aposteriori*, the outer-knowledge which comes as a result of experience.

The *apriori* comprises two parts, *the analytic*, which are statements of pure logic or definitions, and *the synthetic* which have nontrivial content and are generally assertions about the properties of the actual world. Kant's idea was that the synthetic apriori principles exist in our minds even before we have any experience of the world and this provides a framework within which the experience could be understood. A very simple example in this context is as follows; analytic statement *When it is raining one needs an umbrella* may be explained by the synthetic one, because *when it rains water droplets fall from the clouds and the umbrella protects one from getting wet.* As one can understand the two statements are linked so clearly and the knowledge one finds in them is a simple worldly experience which everyone understands from their childhood. For the case of aposteriori one may think of the information one finds out after a given event or a given interaction. Like for example. if one is in a financial service he/she can possibly predict that the economy of a country does depend on its GDP, knowledge he or she could have gained from data mining (derived facts from earlier observations).

While following this methodology the main questions one needs to deal with are what, how, and why. The mode of answering such questions could be different at different levels. While dealing with youngsters, as their experiences are limited, it is first necessary to prepare a framework limited to the age group one is addressing and put in the facts they are normally expected to know in a common day-to-day language. In this context, the most important input in preparing the framework is to add the story of the discoverer of a concept or the inventor of a gadget. The apriori knowledge today is so large that one may take a good portion of it for granted which need not be right. As mentioned above, it is important to recollect the paths that Galileo, Francis Bacon, and Renè Descartes followed which form indeed common basis even today furthered by other eminent and hard-working scientists.

Galileo's pursuits were very significant as he demonstrated how a repeated experiment leads one to a definitive result and new findings. The famous examples in this context are his discovery of the 'laws of simple pendulum' and his experiment with rolling balls along an inclined plane. The story of the pendulum is worth remembering as it illustrates how he used something he knew about and discovered its association to a new phenomenon. It is said that young Galileo, once while visiting the cathedral noticed the lantern hanging from above swinging. As he was already familiar with the regularity of human pulse rate, he started counting the swings of the lantern with his pulse rate and noticed the periodicity. When he went home he suspended a weight to a string made it swinging and repeated the counting with different weights and different lengths of the string. This experiment and the associated observations and inference taken led him to write down the laws of 'simple pendulum' and what followed as developments is well known.

Rolling of balls down inclined planes was another experiment Galileo did to discover the path taken by the balls for the shortest time taken to roll down the inclined plane with different inclinations (the Brachistochrone) and this illustrated the famous principle of least action (bodies take the shortest path while moving from one point to another in a gravitational field, which may not be the straight line). It is important to realize that this experimental observation got fully established with a perfect mathematical result—came to be known as the 'variational principle'(attributed to Fermat in optics, and to Maupertuis, Euler, and Hamilton in mechanics).

It is important to learn that Galileo while teaching a class "Aristotle's mechanics", that seems to claim—'bigger force produces a larger motion'—wanted to test this belief and so took his students to the leaning tower of Pisa and from the top dropped two stones of different size and mass at the same time. When they all noticed that both fell to the ground at the same time they realized that the old belief was not right and this gave a new impetus to 'test all hypothesis' before accepting them. Consequently, this brought a new perspective to learning and his experiment with dropping stones from the leaning tower of Pisa is now considered as the forerunner to the 'torsion balance experiments' of Eotvos (1885) concerning the equivalence of the gravitational and inertial mass of a body. This experimental fact further lead to the 'principle of equivalence', the basic premise of the General Theory of Relativity of Einstein (1914). [The principle of equivalence asserts that the inertial mass (mass responsible for resisting motion) and the gravitational mass (responsible for attracting other bodies) are equivalent, meaning that bodies of different masses like the two stones mentioned attain the same acceleration in a gravitational field. Eötvos proved this by using a torsion balance to an accuracy of one part in a hundred million (10^{-8}), while later experimenters, Dicke et al. raised the accuracy to one part in a hundred billion (10^{-11}) (Roll et al., 1964). Galileo used mathematics to describe the motion of objects after testing the associated hypothesis experimentally. On learning about the invention of the telescope (1609) he realized the importance of using a telescope in the defence of his country (by watching the distant ships which could attack Venice). Looking at the sky with a telescope and observing the motion of celestial bodies, apart from identifying the four (primary) moons of the planet Jupiter, he observed craters on the moon of the Earth and also constellations of stars of the Milky Way. All these observations and the associated mathematics convinced him to support the heliocentric theory of Copernicus as against the Aristotelian view of the geocentric universe. Unfortunately, Galileo had to pay with his life for disproving the unsubstantiated old beliefs as he was restrained to house arrest. Future generations who understood his scientific methods and immense contributions to the understanding of Nature, consider him as instrumental in introducing the objective method for doing science.

Francis Bacon almost a contemporary of Galileo was not considered a scientist, though he was the one who enunciated unambiguously that one should break free of the earlier beliefs and go by actual observations or experimentation. He recommended observing nature critically as well as collectively. He emphasized both inductive logic (going from particular to general conclusions) and deductive logic (going from general to particular) as being necessary for ascertaining the truth of any statement. His main outlook of suspecting and being sceptical about statements before examining all aspects of any proposed axiom was new but lead to debates that were very healthy for the progress of science. He was supposed to be a great admirer of books and a patron of libraries. Talking about books he seems to have said *"Some books are to be tasted; others swallowed; and some few to be chewed and digested."*[1]

---

[1](1) "Francis Bacon." Wikipedia, The Free Encyclopedia. Wikipedia, The Free Encyclopedia, 15 Nov. 2021. Web. 15 Nov. 2021.) also 2. Murray, Stuart (2009). The library: an illustrated history. Nicholas A. Basbanes, American Library Association. New York, New York. ISBN 978-1-60239-706-4. OCLC 277203534.

Humanitarian outlook influenced his work and led him to practice science for the betterment of mankind as he emphasized that the ultimate purpose of doing science is to find avenues to enrich the human life. He recommended collective approach to look for a solution to any problem as more observations of any chosen phenomenon would lead to a more reasonable and believable answer. Modern scientific research indeed follows this pattern quite effectively. In Bacon's praise, Farrington (Farrington, 1951), a noted Irish scholar and a professor of Classics seems to have said, "That knowledge should bring its fruits in practice, that science should be applicable to industry, that men have the sacred duty to organize themselves to improve and transform the conditions of life" (Genta & Riberi, 2019; Farrington, 1951).

Renè Descartes the French philosopher though was a learned man by any standards, always doubted whether all that he had learned was true. He wrote, "*I entirely abandon the study of letters, resolving to seek no knowledge other than that of which could be found in myself or in the great book of the world by travelling and interacting with everyone, gathering various experiences, testing myself in situations which fortune offered me and at all times reflecting upon whatever came my way so as to derive some profit from it*" (BSS, 2017). It has been perceived that he was the first thinker to emphasize the use of reason to develop the natural sciences. He understood the importance of an analytical approach and the importance of asking the right questions. His methodology suggests that one should start from premises that one knows to be positively true and then apply deductive logic through mathematics. The most famous example of his approach lies in his development of coordinate geometry which made mathematics amenable to physical reasoning. The Cartesian system (named after him) indeed brought together algebra and geometry, which were otherwise considered as disconnected.

The wealth of knowledge created through time by the contributions of thinkers who have sacrificed their possessions and comforts and worked hard to attain that knowledge is noteworthy. It has formed the basis for understandings of nature that we have today and derive benefit from. It is useful and important to know about the circumstances under which they worked and created knowledge upon which humanity is progressing.

Anecdotes about the men and women of science, provide a conducive atmosphere for children to be attentive and take notice of what follows after the story. As the story of 'Eureka' is well known, it is useful to recount the background. Archimedes had the problem of finding out the cheating done by the jeweller regarding the mixing of inferior metal with gold in the crown. When he noticed the water overflowing from his bathtub, he must have realized that 'the amount of water which overflew must have a relation to the body immersed', and that he had an answer to the problem. As he knew about the density and its variation for different metals, he realized that crowns of different metals would displace different amounts of water when immersed, and thus using this knowledge he proved the theft of the jeweller to the king. Very often one tends to think about great scientists as genii implying as though genius is someone born so and thus privileged. It is not so! As Thomas Alva Edison the inventor has said, 'Genius is one percent inspiration and ninety-nine percent perspiration'. Creativity is one of the hallmarks of a scientist, and practicing science is indeed an attempt to create a new understanding using the existing knowledge. In his essays on 'The structure of scientific revolutions, Thomas Kuhn (Kuhn, 1962), while discussing the role of history in science says, "If science is the constellation of facts theories and methods collected in current text books, then scientists are the men who successfully or not have striven to contribute one or another element to that particular constellation. Scientific development becomes the piecemeal process by which these items have been added, singly and in combination, to the ever growing stockpile that constitutes scientific techniques and knowledge".

Chroniclers of science have realized that the concept of development by 'accumulation of already known facts' makes it harder to answer questions like 'where was oxygen discovered? Who was the first to conceive conservation of energy? and such questions. There has also been a feeling that it is incorrect to ask such questions as science does not develop by the accumulation of individual discoveries and inventions. It is now well known that Aristotelian Universe or Galen's Anatomy, are not fully correct but at the time when they were current, were no less scientific. Obviously, science has included books of belief quite incompatible with the postulates we hold today. In a strict sense out of date theories are not unscientific just because they have been discarded. Scientific development is a process of acquiring knowledge through the repeated process of falsification of the old and validation of the new knowledge. J. Bronowski defines *science as the organisation of our knowledge in such a way that it commands more of the hidden potential in nature. Man masters nature not by force but by understanding through science.'*

The scientific methodology cannot be and need not be compartmentalized. Instructed to examine electrical or chemical effects of a phenomenon, the man ignorant about these fields, but who knows what it is to be scientific, may legitimately arrive at numerous incompatible conclusions. Among these, several would probably be the ones derived from his own experience or from the knowledge he had gained from investigations by others. What personal knowledge and training does one bring to the study of chemistry or the electrical property of phenomena? Which of his/her hypotheses can lead to scientific development? As expressed by Kuhn, what experiments would he/she choose to perform? What aspects of the complex phenomena would then strike him/her as particularly relevant for his/her investigation? Answers to these questions are the ones where observation and experience restrict the range of admissible scientific explanations. However, an arbitrary element compounded through personal experience or historical accident can always be the formative ingredient of theories offered by a scientific community at a given time. Does effective research begin before a scientific community thinks that it has acquired firm answers to questions like what are the fundamental entities of which the Universe is composed? How do these interact with each other and with our senses? What questions may legitimately be asked about such entities and what techniques may and can be employed in seeking a solution.

In a developed scientific community, answers to such questions come from a firm educational background that prepares and encourages the students for a professional approach that can train him/her as a scientist. Only when the basic education has been rigorous and inquisitive does any new finding will lead a scientific mind to efficient research activity. This makes the research a strenuous and devoted attempt to understand 'Nature' as a part of one's conceptual framework and also helps the person to fit into the framework that can lead to further progress. Guidance by the experienced is also an integral part of such an education.

Science, as practiced today, the activity in which most scientists inevitably spend almost all their time, works on the assumption that the scientific community knows what the world is like. As much of the success derives from the community, willingness to defend accepted assumptions (at times with considerable cost) may often restrict fundamental and novel ideas in the fear that such a move could be not 'in line with the community's basic thinking framework. However despite impediments, so long as the frameworks retain an element of flexibility, the very nature of research ensures that new ideas cannot be suppressed for very long.

Sometimes a research question, one that ought to be solved by known rules and procedures can overcome the repeated criticisms of experts with similar experience. But there could be times when a piece of equipment designed and constructed for research or a set of premises does not perform in an anticipated manner, and the progress of science may get compromised. Under such conditions, when the profession can no longer identify anomalies that question contemporary scientific practice, the advancement may stall. Such impediments can call for scientific revolutions through novel investigation leading to a new framework for doing science. Such scientific revolutions were provided by several eminent scientists like Copernicus, Galileo, Newton, Lavosier, Darwin, Dalton, and Einstein.

The methodology of science evolves with every new phenomenon and its possible explanation. As mentioned already, the basic approach for science consists of hypothesizing, questioning, experimenting, reasoning, and drawing logical conclusions with multiple validations. Good conclusions may often lead to new questions, taking one to a deeper understanding of the phenomena and the principles that govern them. Scientific reasoning leads to and is led by a continued process of thinking. Talking of hypothesis, it is useful to quote W.I.B Beveridge's illustration of the comparison of involved research with Columbus's discovery of the then called 'new world'. (a) Columbus was obsessed with an idea that—since the world is round he could reach the orient by sailing west, (b) which apparently was not original but he had obtained some additional evidence, and (c) he met great difficulties in getting someone to provide money to test his idea as well as in the actual carrying out of the experimental voyage (d) when finally he succeeded he did not find the expected new route but instead found a whole new world (e) despite all evidence to the contrary he clung to the bitter end to his hypothesis and believed that he had found the route to the orient (f) he got little credit during his lifetime and neither he nor others realized fully implications of his discovery (g) since his time evidence has been brought forward showing that he was by no means the first European to reach America (Beveridge, 1957). This has a lot of parallels in the adventures of scientific investigations and in a sense should be a model for scientific methodology except for continuing to believe in the old hypothesis irrespective of the new evidence. Intuition is an important feature of scientific methodology. Of course, this is something one either has naturally or develops through a repeated process of deduction and induction with logical reasoning. Here as soon as some facts are given a scientist feels a certain amount of conviction and a possible hypothesis to explain the observations. Many of the mathematical developments have been due to intuitive thinking and the most famous example is that of Srinivasa Ramanujan and his discoveries with numbers and related functions. Any person endowed with a good measure of self-criticism and intuition generally can gain a deeper understanding of any newer phenomena. Very often intuition is referred to as 'the gut feeling' and following without proper logic can lead to misunderstanding. Putting it strongly, a scientific way of thinking is a way of keeping oneself away from being fooled by others or even by oneself. One of the most significant aspects of science is its universality. This is best illustrated through the fact that there is no state of absolute rest in the Universe. Every experiment performed, or observation made from the Earth is being done at different times in different locations, simply due to the relative motion of the Earth around the sun, of the sun in the galaxy, and of the galaxy in the cluster of galaxies and so on as motion is an integral part of existence. This is why the physical constants used while explaining a phenomenon has universal validity, with every event of observation being made at different points in space-time, and matter that encompasses the Universe.

An important aspect of science that needs emphasis is the difference between conviction and dogma. A proper understanding of the basic structure of a theory along with its fundamental limitations would help in developing a certain amount of conviction. This is important and necessary for putting forth new observational features to be expected or to be interpreted. But this should not lead one to make his/her beliefs into a dogma. Unfortunately, there have been cases when even established scientists have made mistakes because of their personal beliefs, a famous example being the controversy between Sir Arthur Eddington and Subramaniam Chandrasekhar (the Indian/American astrophysicist and Nobel laureate), which delayed the prediction of the existence of black holes by almost 30 years. [It may be useful to recollect the controversy purely with the aim not to have any such pre-determined fixation on the behavior of Nature. Chandrasekhar went to Cambridge as a young student in the year 1930 and already on his way to England had worked out the basics of the equations of state of degenerate matter which was believed to be the state of the core of 'White dwarfs' the dead star. Already in Cambridge, the structure and stability of white dwarfs was a hot topic of research amongst the astrophysicists Eddington, Fowler, Anderson, Stoermer who all had thought that the equation of state of matter inside the core of white dwarfs is non-relativistic. On the contrary, Chandrasekhar had argued that as the degenerate matter would be of very high density and a high temperature it has to be relativistic. Applying the relevant statistics (Fermi-Dirac statistics for the free electron gas which is fermionic matter) he had obtained a 'limiting mass' constraint for white dwarfs for their stability. But alas! The then high priest of astronomy—Sir Arthur Eddington completely disagreed with Chandrasekhar's analysis and had ridiculed him in open meetings of the Royal astronomical society. However, Chandrasekhar being extremely confident of his logic and the analysis, published his result on the limit for the mass of white dwarfs (which is 1.4 times the solar mass) which was later referred to as 'Chandrasekhar limit' and eventually earned him the Nobel prize in 1983. If only Eddington had not stuck to his old beliefs regarding stellar evolution, along with Chandrasekhar he could have predicted the possibility of a continued collapse of massive stars to finally end up as black holes.] There could be some other examples too where famous professors could have discouraged students regarding some unexpected results of investigations (of theoretical nature) which might have led to loss of good careers in science. It is extremely important to be open-minded while trying to understand different facets of Nature and all new ideas that may occur during discussions must be analyzed carefully so that nothing important is overlooked or missed.

The scientific method comprises a set of procedures used for obtaining new knowledge about Nature. And this relies on hypothesis, induction, and deduction. A scientist always starts with facts gathered from experience or observations, by logically following the events that could have happened and then makes a set of probable hypotheses or a theoretical model to explain the experience or observation. This is based on his/her understanding through intuition and deduction. Having made the hypothesis, the next important step is verification or confirmation. The same set of deductions may be explained by varied hypotheses and the one that leads to verification is the correct choice that could bring new knowledge. It is useful to recall Bronowski (1958) as he asserts *"No scientific theory is a collection of facts and it will not do to call a theory true or false in the simple sense in which every fact is either so or not so. All science is the search for unity in hidden likenesses and the scientist looks for order in the appearances of nature by exploring such likenesses and this order must be discovered and, in a deep sense, it must be created."* As he further points out a clear example of such a methodology seems to have been Newton's revelation about gravity. Apparently according to Newton's own recollection, due to the onset of plague in Cambridge in 1665, Newton spent time in his native village relaxing in the apple garden noticed an apple fall. It seems what stuck Newton at that sight was the *conjecture* that the same force which reaches the top of the tree might go beyond in space endlessly to moon. With that thought he calculated the force from earth which can keep the moon in its orbit and compared it to the strength of the force at the top of the tree and found that the forces so calculated very nearly agreed. Newton traced in them two expressions of a single concept-gravity. Combining the two so far unrelated events was his creation and thus discovering the force of universal gravity. As Bronowski says that combining of likely happenings and creating something new is progress in science. As he compares similar situations occurred with Maxwell regarding combining electricity and magnetism with light and Einstein combining Time and space with mass and energy and linking them with bending of light by gravity.

Even after a certain scientific theory is verified, scientists would try to punch holes in it by repeated questioning (scepticism as advised by F. Bacon). Any theory that does not permit being questioned is not very interesting as it would not allow for any further development. No theory explains all the facts within a given set of fundamental laws. This indeed is very good as the understanding of the Universe is an unending task that keeps the creativity and innovativeness of the human mind active.

Talking on scientific investigations, biologist Thomas Huxley, says:

> wherever there are complex masses of phenomena to be inquired into, be they of common daily occurrence or be they of more abstruse problems confronting philosophy, the course of unravelling the complex chain of phenomena with a view to getting its cause needs putting forth hypothesis. One must place before oneself a few more or less likely suppositions respecting that cause, and then endeavor to either demonstrate the validity of the hypothesis or reject it altogether by testing in three ways. Firstly, we must be prepared to prove that the supposed causes of the phenomena exist in nature usually referred to as '*verae causae*'. As a second step we should show that the assumed causes are competent to produce the phenomena we are seeking to explain and finally we ought to be able to show that there are no other causes that are competent to produce these phenomena. If we can succeed in satisfying all these three conditions, we shall have proved our hypothesis to the extent possible with the proviso that any additional knowledge could make us modify it.

Thomas Kuhn points out that some of the classic books of science like Aristotle's Physica, Ptolemy's Almagest, Newton's Principia, Franklin's Electricity, and others, did serve as legitimate sources for problems and methods for science practitioners for they presented achievements sufficiently new and unprecedented to the then practiced scientific activity. Further, they were open-ended enough to leave all sorts of problems for the groups of future investigators to follow. Kuhn defined a new term 'paradigm', for concepts that shared these two qualities. The word paradigm is now referred to most of the new approaches in all fields of research, be it in science or any other discipline of human thinking which initiates discussion based on facts as learned from earlier experiences and knowledge in all forms of human pursuit.

In summary, a few brief suggestions are enumerated below for adopting pursuable methodology for learning and practicing science. These are indeed applicable for any branch of study and human endeavor seeking knowledge.

- Identify a fundamental problem of interest in an existing paradigm that draws your attention and occupies your thoughts. As there are different fields of pursuit in principle, everyone will have some special interest learned through formal education or by keen observation. One would like to investigate some particular aspect of a theory or an effect. This needs to be specifically identified.
- Carry out a literature survey on the topic to identify already known facts and results in the particular field. Obviously in the past several experts could have written research publications/books on that same or similar topic. Visiting a library or accessing web resources makes it simpler to carry out a literature search for publications of relevance on the chosen, as well as related topics. One should spend a good deal of time reading and working out the details of the published material.
- Keep a record of the facts and premises that led to the existing understanding of the problem particular to the chosen paradigm. This is important and necessary before one starts attempting in choosing a problem to solve. If one can identify the unresolved issues within the existing framework one has already initiated to look at the why and how of the issue. That is the best and safe approach.

- For experimental research next comes the practical issue of setting oneself with the necessary equipment/laboratory for experiments and/or observations. For theoretical work, the mathematical framework for the analysis of the problem should be formulated.
- Before setting out one's investigations, it is necessary to increase one's understanding of the chosen topic through further reading and discussing with colleagues and superiors. With today's electronic communication this has become much simpler as one does find a lot of material on the web.
- Whenever and wherever possible one should discuss the findings with colleagues and peers and take their comments/critique objectively, and not hesitate to ask or refer to the seniors on the subject at any time during the study.
- As there is always a chance of stumbling upon something completely new and unexpected, one should not discard any fact that one may unravel which may look illogical at that moment. There could be a chance of discovery awaiting. Continued study of the phenomenon (or discussion with other experts) may lead to new understanding. Alexander Fleming's discovery of penicillin is a very good example of an unexpected discovery. The discovery of the usage of fire by the early man was also serendipitous.
- Repeating the procedure to ensure the correctness and reproducibility of the results is very important. Only then should one arrive at conclusions after convincing oneself of the logic and soundness of the explanations and their consequences.
- Finally, one must make sure to announce one's new findings to the scientific community through the well-laid out process of publication. While doing so, due acknowledgment of the sources of the information used or referred to should be made (citing others) and comments and critiques are to be invited for the progress of science as well as one's further learning.
- It is very important to keep track of similar work done by others in one's field of study and consult time to time on their progress as compared to yours, keeping in mind that one should not get depressed or compromise on his/her efforts as compared to that of others.

In fact, before the twentieth century, many of the discoveries came from individuals working in solitude which perhaps was essential for peaceful reflections on the ideas they worked upon. However, both the increasing costs and the rigors of efforts required consistent teamwork post-1900 which was indeed adopted by most of the investigators particularly in the experimental and observational studies. As one of the methodologies one may say that in the present era, science seems to develop faster through collaboration between researchers investigating similar problems. With the advent of electronic and digital communication like email, Google meets (Zoom, Duo, Skype), and archives, the interaction between distant collaborations, discussions, and advice are easy to obtain which should be used optimally. This facility particularly helps groups who can exchange their data and look for solutions jointly in near real-time. The so-called 'Big science' is no longer the effort of individuals working in isolated laboratories but a joint venture as seen in the case of 'the Large Hadron collider' at CERN which identified the elusive 'Higgs' particle, the 'international neutrino program' (INO) that aims to understand the role of neutrinos in the history of our Universe, the 'LIGO-Virgo project for the detection of gravitational waves on earth' which consequently uncovered the 'multi-frequency astronomy as a result of observation of gravitational waves from neutron star collisions, and the 'Human Genome program' which looked for sequencing and mapping of the genome and gave the ability for the first time to read the complete genetic blueprint of the human being. All these programs are directly under inter-national collaborations where the required resources (money and manpower) are shared by many countries and the investigators are in constant link through the electronic and digital communication networks mentioned above.

Before initiation of one's research programs, the relevance of the investigation can be ascertained through the help of Google Scholar/archives/Scopus or the like, which provide a complete list of investigations on any particular subject to date. This information can guide the researcher about the changes he/she needs to make in his/her investigation and emphasize the necessity if required to modify. As men-tioned the methodology of doing science also evolves with every new phenomenon observed and every new explanation sought because of the associated understanding of related aspects.

References

Beveridge, W. I. B. (1957). *The art of scientific investigations*. Norton.

BSS. (2017). *A brief history of Science*. Breakthrough Science Society.

Farrington, B. (1951). *Francis Bacon: Philosopher of industrial science*. Lawrence and Wishart.

Genta, G., & Riberi, P. (2019). *Technology and the growth of civilisation*. Springer.

Iyengar, R. N. (2016). Ancient astronomy in Vedic texts. In K. Ramasubramanian, A. Sule, & M. Vahia (Eds.), *History of Indian Astronomy-a Handbook*. Science and Heritage Initiative. ISBN: 978-81-923111-9-7 pdf. https://www.researchgate.net/publication/30 9710402_

Kochhar, R., & Narlikar, J. (1995). *Astronomy in India: A perspective*. INSA.

Roll, Krotokov, & Dicke. (1964). *Annals of Physics, 26*, 442–517.

Shaver, P. (2018). *The rise of science*. Springer.

Chapter 4
Language of Science

Introduction

The language of Science essentially is a discussion of the way science is communicated to others. While there is an implication of the audio and visual communication using the media of talking, reading, and writing, there is a more important aspect of conveying the concepts and experiences which need to be understood by the receiver. As communication also depends on the age groups, it becomes necessary to keep in mind the familiarity of the various individuals and their capacity to follow the language used. However, scientific principles and laws of nature should be so communicated that the receiver can repeat an experiment or derive a formula irrespective of the geographical location or time of an experiment performed and get the same result. Further one should also bear in mind the fact that just giving a whole lot of information does not convey science. Only when the receiver can experience the joy of understanding the communicated idea, the language can be said to be proper and effective.

Having learned the methodology of doing science one should then think about communicating what was learned. What should be the Language of Science? Language is a medium for communication which can be through varied sensory organs. Generally one uses the terminology language mostly for acoustic or audio communication which is through speaking and hearing. Though, they are the most common forms of communication one knows that experiences can also be had and conveyed through visual and feeling aids such as the use of braille. There is a saying that seeing is believing but that may not be always necessarily true!

As Science deals with the knowledge of Nature its communication could and should be in and with all possible forms of perceptions that are possible. It is indeed relevant that in the educational system there has been considerable debate on the language of teaching. The point being whether it should be the mother tongue/regional language or English that has been considered universal. Science being an experience, the choice of oral language should not matter too much. It is known that a proper understanding is obtained only when one can experience a description. For example, a visually impaired person can hardly appreciate the beauty of a rainbow or a hearing-impaired person the call of a cuckoo.

Communicating science can be a simple and enjoyable pastime provided one understands and experiences the phenomena through one's sensory organs. As science is universal the language in which it is expressed should be adoptable without geographical barriers. Once a phenomenon is described or a law stated, it should produce the same effect or give the same result irrespective of when or where the experiment is conducted or observation is made and who does it. Then only it can be termed as scientifically valid—a universal truth. The most important aspect required for doing and communicating science is thinking. It may appear redundant (strange) if one asks the question 'does thinking need a language'. The answer is yes as the process of thinking helps 'formulation of logical communication within the mind and associate it with words to express the idea that occurred'.

Albert Einstein, the icon of modern science asked: *"What is it that brings about such an intimate connection between language and thinking? Is there no thinking without the use of language, namely in concepts and concept combinations for which words need not necessarily come to mind? Has not every one of us struggled for words although the connection between 'things' was already clear?"*

One may conclude that the mental development of an individual and his/her way of presenting concepts depend to a high degree upon the language, which shows a direct link between language and thinking. This feature can be experienced when one hears the narration of the same events as described by a child and an adult. It is likely that with maturity an adult can follow and explain various steps of his/her experience much more clearly than a youngster with limited experience. This feature can certainly influence their respective mode of communications of the event. While describing an event or an occurrence many times one tries to think of past experiences under similar situations before giving a final expression for the description through words. This becomes clear particularly while teaching (making others learn). Teaching children because of their limited experience can be trickier than teaching grownups. Also, most forms of communication get clearer only interactively which also requires giving expression to one's thinking and usage of words during discussions. It is also possible that while describing an occurrence, a child could be more hesitant due to 'lack of words' than an adult who could have done it earlier and thus will know how to express it in clearer terms.

It is often recognized that one's thinking is incomplete unless one can express the results of thoughts in a communicable way to others. It is important to develop different ways and means of communication which can convey the processes and outcomes of one's thinking to groups of different ages/experiences. Such successful communication should be the language of science. Thus the language of science can be defined as the capacity to explain observed events to others while keeping in mind the maturity and experiences of the receiver under different circumstances and their consequences.

Knowledge and Information

It is also important to distinguish between knowledge and information while trying to communicate. In today's world, digital media enables one to easily accumulate a lot of information on almost every issue of interest. The mere gathering of information and storing it in memory does not create knowledge. Knowledge is the synthesized information that helps in the creation of new knowledge and this needs creativity. Creativity in science is no different from creativity in any other field of activity except for the language and mode of communication/expression. Communicating science needs to be looked at different levels and particularly concerning the general public or groups of educated individuals. Here one needs to differentiate between just giving information or making them understand a new feature of a discovery. Most of the time supplying information to the public on new developments happen through what is referred to as 'science journalism' which also needs to keep in mind the language that can be persuasive. Michael F. Weigold (2001) writing on the topic 'Communicating Science' suggests that "science' refers to a broad range of activities, (a) seeking knowledge for its own sake (basic), (b) exploring solutions to varied problems and concerns (applied), or (c) use their knowledge technologically to develop new products. While media has its interpretation of what science stories are, and it has varied over the years, the public interpretation of the same story may range over as being political or economic or just a technological advance. According to Shortland and Gregory (1991) 'science coverage may have reached its zenith during the second world war when science and technology were seen as integral to victory, and later the launching of Sputnik by the USSR led to renewed interest in science generally'. According to Bader (1990), newspapers that carry regular science sections as compared to those that do not also give greater coverage to science in the news section particularly on basic research. Apart from the journalistic coverage of science, general audience science books also play an important role in popularizing science and making it reachable to different levels of audience. As emphasized by Weigold the internet (WWW) seem to be making a great impact on science communication at least for four reasons (a) scientists and organizations can communicate directly with the audience and no mediation is necessary (b) no restriction of space and time which is inherent in ordinary news media (c) Web combines the information richness of print with the demonstration power of broadcast in a seamless, accessible, interactive fashion and finally (d) Web is an instantaneous two-way communication medium allowing all possible communications between the source and receivers and between receivers. It is useful to consider what Valenti (1999) cautions as regards the difference in the perspectives of a scientist vis-a-vis a journalist regarding informing the public on science. Scientists, being specialists involved with a very specialized aspect of a larger puzzle, sometimes may not be able to put their part into a story that the general public can appreciate. Journalists may view a scientist as obscure and narrowly focused as he/she looks for stories of exciting potential that may even lead to a controversy but might increase the readership. Scientists offer predictions that are tentative and qualified, which may seem incompatible with fostering excitement in a story, as many times the importance of scientific work is not immediately obvious. It happens that in many cases discoveries could only be an incremental part of a larger undertaking. Of late, most of the scientific organizations have professionals generally called public relations officers, who are supposed to be experts in the language of communication (preferably without jargon) and are capable of linking the working scientists with media persons so that the information shared is authentic and news-worthy. Science by definition has to be objective requiring tests that permit theoretically incompatible outcomes or tests of hypotheses that must be replicable, whereas journalism often needs to be subjective to uphold what is termed as fairness. Science communication today is certainly far easier than what it used to be some time ago, as digital media provides a fair amount of facilities like pictures, models, and demonstrative experiments that could be related to several branches of modern science.

Galileo discovered the fundamental truth about motion and dynamics but what is remarkable are the methods he devised to discover and make these truths realized by other practitioners of science. He introduced mathematics in describing physics! The most important aspect is that anyone could repeat Galileo's experiments and get the same result. This methodology in the modern world has produced several applications (examples will be cited at appropriate places in coming chapters), where one seeks to understand a given phenomenon through several repetitions using different techniques but keeping the final aim undisturbed.

The era of Galileo was followed by that of Newton and the language of mathematics became the basic entity for science. It is well known that Newton discovered the language of calculus (using the infinitesimals and limits) to define the concepts of velocity, acceleration, and momentum to describe the laws of motion and of gravity. Along with mathematics, scientific expression needs logic which sets out either with axioms or experience. One uses several terminologies while presenting scientific facts; _viz_ axiom, postulate, principle, conjecture, lemma, theorem, theory, and models. The meaning and the usage style of these terminologies are briefly discussed below. Though mathematics as a communicating language could be difficult for telling stories, the terminologies explained below are all very common to define and express the principles and theories in most of the branches of science.

Axioms are assumptions that are set out as basic premises from where logical deductions explain certain definitions and phenomena. Axioms are not to be questioned. They are also defined as self-evident statements! For example, _'a point has no dimension'_. This is the first statement one learns before expressing the facts of Euclidean geometry. Axioms, referred to as self-evident truth in common practice, refer to some of the beliefs that one has taken for granted but seem logical. This calls for due care while setting up an axiom. For example in the statement above defining a point, the definition of dimension of space pre-empts the definition of a point. The usage of an axiom as self-evident truth extends beyond the practice of science, as many of the human activities assume some basic knowledge of events that could follow their actions and reactions. As suggested by Courant and Robins (1969) the axiomatic point of view can be as follows: To prove a theorem in a deductive system, one shows that it is a necessary logical consequence of some previously proved proposition, which in turn also was a consequence of another similarly proved result. As this procedure of deductions cannot be continued forever at some stage one is permitted to stop and allowed to assume the statement as an accepted truth and called it an axiom. If the facts of a scientific field are brought into such a logical order that all can be shown to follow from a selected number of statements then the field is said to be presented in an 'axiomatic form'. Hilbert is credited with the contribution that he gave a satisfactory set of axioms for geometry and also made an exhaustive study of their mutual independence, consistency, and completeness.(Courant & Robins, 1941).

Postulates are similar to axioms but could be the result of reasoning and form a basis to arrive at new interpretations of observed facts or logical deductions. Postulates are proved wrong or incorrect when the result based on them do not hold good. Failure of a postulate opens up new avenues of a search for truth as one then seeks to develop an entirely new process of thinking, the consequence of which could be a new advancement. The best example of such a possibility is the following.

Consider, for example, the 'parallel postulate' of the Euclidean geometry (plane geometry) which states that, "*on a plane, for a given point and a line, one and only one straight line can be drawn that passes through the point and is parallel to the given line*". This is the basis for the entire Euclidean geometry. Assuming that this need not hold, i.e., no line parallel to the given line is possible to be drawn, Riemann (German mathematician, student of Gauss) developed the spherical geometry, which holds on surfaces with positive curvature (like a sphere). On the other hand, Lobachevski (Russian mathematician) assuming that more than one parallel line can be drawn, arrived at the hyperbolic geometry with negative curvature (an example of a surface with negative curvature is a saddle surface.) Riemannian geometry was a mere mathematical development till it was used by Einstein (on the advice of Marcel Grossman) to formulate the theory of general relativity which could not be described on a flat geometry like that of special relativity, and where gravitation is identified with the curvature of space-time (Prasanna, 2008). There is an important aspect of proving a postulate by a method called 'reductio-ad absurdum' wherein, one first assumes a given statement as not being true and derives from that assumption a logical result that is known to be wrong. Such a contradiction naturally upholds the opposite point of view supporting the statement, and proves the postulate to hold.

Principles are foundations on which scientific theories are based and built. These could be the results from experiments or defined through mathematical deductions. Archimedes' principle is the earliest example of a scientific principle that was realized by a keen observation by the discoverer. The story of 'Eureka' is well documented and children learn it at school. This principle laid one of the basic foundations for the science of fluids, fluid mechanics. Almost all aspects of scientific investigations are normally based on one or the other principle. To name a few, classical physics is based on Galileo's principle of relativity and the invariance of Newtonian mechanics (true for all observers in uniform motion). Quantum physics is based on Heisenberg's uncertainty principle, and Pauli's exclusion principle.

Similarly, the entire description of the structure of the Universe is built on the Cosmological principle which assumes that the Universe is homogeneous and isotropic. The thermodynamic principle of energy conservation and the fact that energy conversion from one form to another is not always fully efficient have their imprint both on physical and biological sciences and are equally important.

Lemmas and theorems are mathematical entities, which can be proved, using either axioms or postulates. Euclid's Elements, the book on plane geometry has theorems on the properties of triangles and circles which are basic to several aspects of engineering drawings (for example Pythagoras theorem), and graphic designs. Theorems can also be statements of laws of Nature that can be tested experimentally or proved theoretically (for example laws of thermodynamics).

Conjectures, Theories, and Models are scientific facts that can be verified either through mathematical logic or by experiments or observations. Conjectures are 'statements of inferred principles' that reflect from results of experiments or observations. Sometimes these could be inconclusive but can be of good use in helping to design/plan experiments/observations for a new set of physical parameters which eventually through a feedback process lead to new understandings and thereby to new theories. Most conjectures are mathematical in their formulation. Some famous examples are the (1) principle of least action [attributed to Maupertuis and Euler], which is a principle that states 'all actions of Nature are such that the energy spent on taking the action is the one along which it is minimum of all possible paths'(like for example on an Euclidean surface the straight line is the shortest path between any two points while on a curved surface it is the geodesic). (2) Einstein's principle of equivalence [already mentioned in Chap. 3] which indeed is one of the basic requirements of the theory of general relativity, (3) Fermat's last theorem [the only integer solutions that can exist for the equation, $a^n + b^n = c^n$, where a, b, c are positive integers is for the case $n \leq 2$ (proved by Andrew Wiles in 1993). In the case when $n = 2$, the equation with equality describes the Pythagoras theorem with a and b being the sides of a right angled triangle and c its hypotenuse]. It is worth mentioning that the enigmatic Indian mathematician Srinivasa Ramanujan's work and contributions to mathematics were mostly intuitive conjectures, and several mathematicians have spent years of research to prove many of his conjectures. Conjectures are an integral part of a scientific process. It is always possible that sometimes a conjecture may not hold good, and demonstration of its failure leads to the reanalysis of the hypothesis on which the conjecture was based. The process thus can lead to an improved understanding of the phenomena or helps in developing new lines of research.

Theories are explanations of laws learned through observations or theoretical analysis and generally should satisfy the criteria that they explain some of the known results and also predict new effects, which can then be verified. The famous examples in this context are the Newtonian theory of gravitation, Dalton's atomic theory, and Einstein's theory of general relativity explaining Gravity as the theory of space, time, and matter and the more prominent Quantum theory of Planck, developed by Bohr, Schrodinger, Heisenberg, Pauli and Dirac which is at the basis of the twentieth-century physics along with the special theory of relativity.

Newtonian theory of gravity, not only explained why things fall to the ground, but also the motion of celestial bodies, particularly that of planets around the Sun, and of moon around the Earth. Kepler's laws of planetary orbits are completely consistent with Newton's laws of motion and the law for gravitational attraction between bodies. The success of space technology is mainly because one can predict the possible orbits for the satellites by using Newtonian mechanics (today one takes into account the possible corrections due to general relativity) and also can program in advance for possible course corrections that may be required when the satellites are in orbit. The General theory of relativity propounded by Einstein exactly explained the lacuna that was existing between the Newtonian prediction and actual observation as in the case of the perihelion shift in the orbit of planet Mercury around the Sun (approximately 43 secs of arc per century). The theory also predicted that light from the stars passing across the Sun's disc would get deflected by about $1.75''$ and this was tested during the total solar eclipse of 1919 and was verified to be correct. The correctness of Dalton's theory was established by the atomic models of Bohr and Somerfield explaining the atomic spectra and the chemical reactions which have formed a strong basis for the chemistry of matter and further plays a fundamental role in understanding the molecular and atomic bonding required to study the structure of matter. In life sciences, the theories of evolution and natural selection, are well known. It is scientifically interesting and important that these theories are being debated and new evidence is sought for a fuller understanding of life on earth. In fact, over the last few decades, developments in mathematical and computational biology have opened a new phase of research in Biomathematics and the science of Genetics which are expressed as laws and theories, and further studies in these areas can advance our understanding of Nature many folds.

Models, most often are the results of weaving together observed or learned facts about nature and its laws, in terms of mathematical entities, which can yield quantitative results associated with the discussed phenomena. Often a model is referred to as the mathematical relation that describes some real-life situations. These make it easier to test as well as communicate the acquired knowledge in an unambiguous manner. This particular aspect of learning in science has advanced rapidly with the advent of computers and programming languages, as they facilitate the handling of very large numbers and repetitive calculations with ease and speed. Recent rapid progress in sciences could be attributed to mathematical modelling which is used as the basis for exploring (extrapolating or interpolating) the possible consequences or reasons by using the already known facts (data mining i.e. predicting based on known features through computer programs of existing statistical data). The conclusions have been mostly sound, reasonable, and trustworthy. During the recent COVID-19 pandemic, numerous predictions were made about the growth and decay of the virus causing the pandemic using the collected data and their mathematical modelling. Indeed, many of the predictions did not turn out to be accurate as the database were insufficient and the scientific understanding of the virus and knowledge about its mutation and spread was limited. On the positive side however, the predictions and projections made from them helped some understanding of the problem, and data collected by the concerned agencies and their analysis have been very useful for the vaccination program and for taking more precautions. Modelling has played and plays an important role in the financial markets which can influence the economic status of a country or of an individual.

Paradigm: As mentioned in an earlier chapter the new terminology very popular in research is the notion of a paradigm, introduced by Thomas Kuhn (1962). Referring to 'normal science' as research firmly based on earlier scientific achievements, which were recognized by some particular scientific community he defines 'paradigms' as achievements sufficiently unprecedented but open-ended for the future practitioners to solve. He opines that the emergence of such paradigms affects the structure of the group which practices the field but at the same time it could lead to a reformulation of the paradigm supported both by theoretical and experimental results of the latest discoveries. Many theories explaining the fundamental interactions of nature do come under such reformulations particularly the developments in the twentieth century.

The accepted language of science is the description of a logical chain of events verifiable at any point in space and time with either experiments or observations explained with the help of accepted knowledge or with new inputs of significance. Apart from the technical correctness of any report of observation made or experiments performed, the more important aspect one should be aware of is the precision of the expression and the accuracy of measurements used for the description. Teachers often give examples of the errors students generally make while expressing physical quantities associated with experiments concerning units and dimensions. Making an error in expressing a quantity is often compared jokingly to someone answering the question what is your height? With the answer Oh, about five miles! Though this example appears a little too exaggerated, the need to be very careful while expressing the units of physical quantities should not be overlooked. Measurements mentioned for an experiment in chemistry have to be doubly precise as mistakes even in decimal places may result in accidents for someone else trying to do the experiment believing the reported measurements. Thus accuracy is the hallmark of science reporting which may at times be overlooked in the hurry to get priority of publication. The need for peer review of reports and submissions cannot be overemphasized than these examples.

Learning comes through associating the new results with the existing data with appropriate explanations. It is possible that quite often one can explain how an effect happens more clearly than to describe why it is so. This could be because there could exist more than one reason for producing the same effect. In such a situation, the scientist will enlist all the probable reasons and then applies the causes logically possible before coming to a proper conclusion.

Talking of science communication, one of the best methods particularly for the case of young adults and the general public is through the establishment of 'community science centers'. Vikram A Sarabhai, the doyen of space science in India initiated one such venture in 1967 at Ahmedabad, India, which is very popular and useful for communicating science. The most important aspect of the center is to initiate children towards science learning through doing experiments by themselves illustrating scientific principles in various branches of science. A very important aspect of this center is the design and development of mathematical models depicting different properties of geometrical concepts and figures. In the same context science museums have been playing a very important role in educating the public on several aspects of science and technology. Very often there are several competitions and/or exhibitions both at school and college levels which encourage youngsters to learn science communication both verbally and through simple demonstrations of known facts in science. Also to create public awareness and understanding of science, often popular lectures are arranged on topics of special interest by science museums and sometimes even by research institutions.

Science communication became much more effective and communicative with the development of the discipline 'Cybernetics'. Of the various different definitions offered (von Foerster et al. 1951) emphasising it "*as a form of cross disciplinary thought which made it possible for members of many disciplines to communicate with each other easily in a language which all could understand*", seems the most appropriate. Effective use of feedback is very important while practicing and communicating science. The expression 'language which all could understand' may not be simple but is a necessary attribute for explaining scientific or technical issues. Often it is only descriptions heavily worded with jargon that causes discomfort among the listeners and failure of communication and only through getting a feedback one can make the communication effective.

References

Bader, R. G. (1990). How science news sections influence newspaper science coverage: A case study. *Journalism Quarterly, 67*, 88–96.

Courant, & Robins. (1941). *What is mathematics*. Oxford University Press.

Kuhn, T. (1962). *The structure of scientific revolution*. University of Chicago Press.

Prasanna, A. R. (2008). *Space and time to spacetime*. Universities Press.

Shortland, & Gregory. (1991). *Communicating science: A handbook*. Longman.

Valenti, J. (1999). Commentary: How well do scientists communicate with media? *Science Communication, 21*, 172–178.

von Foerster, H., Mead, M., & Teuber, H. L. (Eds.). (1951). Cybernetics: Circular causal and feedback mechanisms in biological and social systems. *Transactions of the seventh conference*. New York: Josiah Macy, Jr. Foundation.

Weigold, M. F. (2001). *Science Communication, 23*(2)164–193.

Chapter 5
Mathematics, the Common Base for Science

Introduction

It is known that the results of experiments and observations are expressed in terms of numerical data coming out of the governing equations and formulae. Mathematics is the language of numbers associated with arithmetical operations and their consequences. It is therefore natural that it stands as a base for expressing the principles and results of investigations in science. Thus one can measure and quantify the results of one's investigations through numbers and compare/contrast them with those of other investigators for verification. A change in the numbers may result either in a new finding or an error in one's investigation. These can generally be checked through logical steps based on the theory supporting the problem under investigation. Mathematics is a beautiful construct of the human mind expressed through numbers, lines, graphs, and equations following logical reasoning. Scientists like Galileo, Newton, and Einstein were the ones who brought in a new and sophisticated approach to scientific culture through the language of mathematics. It is not surprising that the puzzles and riddles often pose a thinking mental pastime and sharpen the mind through the quest for their solutions.

Mathematics: Its Need in Science

The language of science has to be communicable and reachable to all as knowledge about our Universe has been an innate curiosity of every individual. The human mind grasps numbers far more easily than abstract concepts and therefore it becomes desirable to express the results of all investigations in terms of numbers and associated algorithms. Further, all physical laws and natural phenomena are best evaluated quantitatively with numerical data from observations and calculations based on theories associated with abstract concepts. Comparison of different events, be they related to stars in deep space or to microbes within the cell walls have to be carried through quantitative measurements using varied instruments ranging from telescopes to microscopes. As all measurements of events occur in space and time it is necessary to have a direct association of points in space with unique sets of numbers which is done through a coordinate system. Mathematically this is done by defining the notion of a function that identifies points in space with the numbers in a coordinate system. As a set of points define a line or a curve depending upon the mathematical relationship the coordinates satisfy such functional relations help in visualizing the lines and curves as trajectories of events in space-time.

Through the coordinate systems, mathematics provides an apparatus for the language of numbers. Euclidean geometry and Algebra work with several independently introduced concepts and symbols such as the real and complex numbers, the point, the straight line, and operative signs $(=, +, -, \times)$ which designate the fundamental operations that provide connections between the various concepts such as addition, multiplication, fraction, etc. Ordinarily to measure an event one needs to associate numbers (functions) with the location and time of the event. This requires a direct link between the spatial position of the event and the time, which is achieved through a coordinate system. For this one uses three numbers, associated with the three dimensions of space identifying the location in space synchronizing with all our experiences in the world outside and one other number to identify the time of a particular happening (event). Working with such a $(3 + 1)$ dimensional coordinate frame of space and time at an origin of the coordinates one can then measure in terms of the four numbers expressing where and when an event occurred. If the events occur at different locations at different times, these four numbers change and give the history of events which is also called as the world line.

The operations of addition, subtraction, multiplication, and division are represented through familiar symbols from arithmetic. This is the basis for the construction and the definition of all other statements and concepts. One calls this approach coordinate geometry and the coordinates Cartesian coordinates (in honour of *Renè Descartes*) and is a fundamental feature of mathematics as used by scientists and engineers. As the science progressed, extensions and alternates (spherical polar and cylindrical polar coordinates) to this approach were developed. The basic geometry for all these was the flat Euclidean geometry. Later developments lead K F. Gauss to introduce the geometry on curved surfaces which was later extended to n-dimensions by his student Riemann. It is the four-dimensional Riemannian geometry that formed the framework for General Relativity-the extension of Special relativity and Newtonian theory of gravitation, which we will discuss in a later context. The connection between concepts and statements on the one hand, and the observational data on the other is established through counting and measuring well known to all and used in daily life.

In the words of Courant and Robbins (1941), *"Mathematics as an expression of the human mind reflects the active will, the contemplative reason, and the desire of aesthetic perfection. Its basic elements are logic and intuition, analysis and construction, generality, and individuality. Though all mathematical developments have their psychological roots in practical requirements, it invariably gains moments in themselves and transcend the confines of immediate utility*". Explained simply, this statement implies that developments in mathematics comes generally for usage in daily life on one hand, while a more serious and intellectual approach leads to developments that finds usefulness possibly at a much later stage for purposes originally not even thought for.

Evolution of Early Mathematics

Recorded evidence suggests that mathematics started about 2000 BC in the Orient and the Babylonians added a great wealth of material which today goes under the name of elementary algebra (Courant & Robbins, 1941). However, as a science and a tool for science, mathematics emerged in Greece during the fifth and fourth centuries BC. It is said that the Greek thinkers became conscious of the inherent difficulties associated with the concepts of continuity, motion, and infinity, and particularly in the problem of measuring arbitrary quantities by given units. One of their leading philosophers, Eudoxus (already referred to) developed the theory of the geometrical continuum and seems to have clarified some of these concepts. Also, he is the one to have initiated the idea of a geocentric system of the world (as mentioned earlier), and to describe the physical events like the motion of a planet by a set of functions that are regular and smooth (implying continuity). [A function is said to be continuous at a point if it has the same numerical value as one approaches the point from its right or left on the real line. A very simple example of a continuous function is given by the well-known trigonometric functions $Sin[x]$ or $Cos[x]$ in the interval $(-\pi \leq x \leq +\pi)$.] As ideas become clearer, the importance of dealing with continuous functions becomes necessary when one wants to introduce the concept of differentiability (this concept came in with Newton and Leibnitz) that is required to describe the history of any system in time, like planetary orbits.

The development of Mathematics in ancient India was associated with developments in Astronomy as evidenced from the works of Aryabhatta (500 AD), Varahamihira (600 AD), Brhamagupta (700 AD) and Bhaskaracharya (1200 AD). After introducing the digit zero, Indian mathematicians advanced the notion of place value which included the decimal system. The concept of infinity is self-explanatory. By definition, one can express it as what exists beyond all boundaries and is unmeasurable. In simple terms (operationally) one can identify it as 'any finite quantity divided by zero is infinite'. In the ancient Indian context it was defined in Sanskrit as 'Poorna' (fullness) which is described as:,

"*Poornamidam, Poornamadam, Poornath Poornamudachyate.*
Poornasya Poornamadaya, Poorna meva vishishyate."
Meaning, 'Infinity here, Infinity there, Infinity arises from Infinity.
Taking out Infinity from Infinity leaves behind Infinity'.

One of the first structured treatises on geometry was 'Euclid's Elements' which describes theorems concerning parallel lines, triangles, circles, areas, and volumes.

The study of properties of these structures was through a methodology first enunciated by Eudoxes where for a given data, one constructs some helpful pictorial
representation and then proves (verifies) the given statement like for example 'the
sum of the three angles of a triangle is 180°'. Technically such a procedure was
known as the 'deductive–postulation', which attempts to explain a phenomenon
from the data given, by assuming a certain possibility (postulate) and then deducing
(proving) the consequence through logical reasoning.

The introduction of the number system for counting and its utility has different
versions with claims crediting both the Arabs and the ancient Indians, for its
discovery and usage. It was thus termed as the Hindu-Arabic numeral system
[https://en.wikipedia.org/wiki/Numeral_system].

However, it is generally accepted that the decimal system was invented in India as
the place value method and was used by the early creators of Indian knowledge
Aryabhatta, Brahmagupta, and Varahamihira mentioned above BSS (2017).

Mathematics as a Tool for New Discoveries

As cited in (Courant & Robbins, 1941) after a period of slow progress, mathematics
and sciences witnessed a revolution during the seventeenth century with the developments in analytical geometry and the discovery of calculus. In the nineteenth
century, the imminent need for consolidation of what was learned till then and the
desire for more security in respect of extending it to higher learning (prompted by the
French revolution) inevitably led back to a revision of the foundations of new
mathematics, particularly of the differential and integral calculus and the concept
of limit. It is opined that these developments lead to new advances and the realization
of the need for precision and rigorous proof. At this stage (in the nineteenth century)
it was also hoped that the regained internal strength of mathematics and the enormous simplification attained based on clearer comprehension (understanding) made
it possible to master the mathematical theory without losing sight of its applications.
This perhaps led to a useful combination of pure and applied mathematics retaining
both the features; abstract generality and applicable individuality. Expressing more
simply, it is opined that the developments in the science of mathematics during the
last five hundred years, mainly emphasized the need for rigor and precision (abstract)
as well as find its usefulness (applicability) in putting other branches of sciences on a
firm and logical foundation of mathematics.

As science requires expressing concepts and quantification of results with precision it aims to find a basic framework for deducing the laws that govern any natural
phenomena and it does so through the language of mathematical equations
and formulae. As well-formulated explanations obtained through keen observation
and deductive logic are the goals for understanding all laws of Nature, intuition and
construction are the driving forces which mathematics provides. It has been sometimes suggested inappropriately, that 'Mathematics is just a system of conclusions
drawn from definitions and postulates which are consistent'. While this could in

some sense be true operationally, mathematics as a language of science has more to offer all thinking minds. The idea that the mind can create a set of assumptions whenever it wants is not correct. Only under the discipline of responsibility and due consciousness to understand the complete structure, and guided by intrinsic necessity can the free mind achieve truths of scientific value. The breakthroughs in several aspects of science have resulted because of the advent of mathematical concepts to describe physical phenomena and also look for their extensions. A good example of such a development is the description of Gravity, the weakest force of Nature, but yet the most dominant of the natural order! It is not a surprise that the revelations associated with Gravity have come through the most brilliant of the human minds, Newton and Einstein who both used new mathematics to develop their theories as narrated briefly below.

Newton invented calculus-using the notion of infinitesimals (very small increments) and the notion of differentiability through which he could define the concepts of velocity and acceleration precisely. These along with the Euclidean geometry play a fundamental role in deriving the equations of motion (orbits of the planets as Kepler had surmised) and subsequently the inverse square law of gravity. [In this context it is important to mention the contributions of Leibnitz (Germany, 1646–1716) who was the first to introduce the notation for the integral $\int f(x)\, dx$, as also to discover the formula, $d(x^n) = nx^{(n-1)}$, most important result in differential calculus.] Newtonian formulation of Gravity dominated the description of Cosmos and the world for more than two and half centuries until Einstein revealed the truth of the Cosmos as the geometry of Space, Time and Matter, through his General theory of relativity. For describing the theory, Einstein had to learn and adopt the beautiful Riemannian geometry (non-Euclidean geometry) and tensor calculus which was developed earlier by the giants of mathematics—K.F. Gauss, G.F. Riemann, T. Levi Civita, and L. Bianchi. The beauty of the theory of general relativity is entirely due to its underlying mathematics, and the theory was, no doubt, termed as "Geometrisation of Physics" by the American physicist J. A. Wheeler.

It is well known that Einstein used a four-dimensional framework of space-time for elucidating the Special theory of Relativity, as he gave 'time' (no longer considered absolute) the status of a dimension along with the three spatial dimensions. This was necessary as in the new formulation 'time' had lost its unique stature of being the same (absolute) for all observers irrespective of their motion. This lead to the notions of time dilatation (moving clocks go slow) and length contraction (moving rods appear shorter) for a stationary observer. However, this theory describes physics only on a flat space (Minkowski background) and thus the metric (distance between any two points) coefficients are all constants. Most importantly this theory unified the Classical mechanics of Newton and the Electrodynamics of Maxwell and further stood as the basis for the new Quantum mechanics which came about in the 1920s.

When Einstein wanted to extend the special theory to include the accelerated observers, and thus take into account 'gravity', the general theory of relativity could not be accommodated on the flat Minkowski background thus leading him to describe the gravitational field on a curved Riemannian background geometry (non-Euclidean geometry with positive curvature), at the suggestion of the mathematician Marcel Grossmann. Einstein prescribed the equations governing the gravitational field of a material system, now well known as Einstein's field equations bringing together the material distribution (energy-momentum) with the geometry of space-time. The most important feature of the theory was the identification of the gravitational field with the curvature associated with the underlying space-time. Considering the world lines (particle histories) of particles in a gravitational field, it has been shown that the background curvature, represented by the Riemann curvature tensor identifies physically the relative acceleration between the two particles in a gravitational field. As gravity is a long-range force the equations allowed physically relevant solutions even outside the matter distribution. The most important and interesting solutions for these differential equations (solved by Schwarzschild (in 1916) and by Kerr (in 1963)) represent the geometry of non-rotating and rotating black hole space-times respectively. There also exist solutions that describe the field of 'Gravitational waves' which is also a perfect confirmation of Einstein's theory. Beyond these, the theory provides a beautiful and consistent background structure for describing Cosmology the theory of the Universe.

Whereas the special theory of relativity is an absolute must for describing electrodynamics, the phenomena related to charges, currents, and magnetism the general theory is a must for understanding the phenomena associated with gravity on a large scale and for describing our universe. (Prasanna, 2017). (It is worth mentioning that all these results of Einstein's theory of general relativity earned Nobel prizes; in 2017 to KipThorne, B. Barish, and R. Weiss, for Gravitational waves, in 2019 to Jim Peebles for Cosmology, and in 2020 to Roger Penrose, Reinhardt Ghenzel, and Andrea Ghez for Blackholes.)

In the field of life sciences, the role of mathematics is well illustrated from the works of G. Mendel who was among the firsts in making mathematical modelling of his experimental observations for deducing the law of heredity (Teicher, 2014). In the last fifty-odd years' Mathematical biology has been a popularly emerging area leading to well followed scientific research both leading to and led by technological innovations and experimental facilities. R Fischer (1890–1962), J. B. S. Haldane (1892–1964) and S. Wright (1889–1988) were responsible for combining Mendelian genetics and Darwin's theory of natural selection to evolve the neo-Darwinian theory of synthesis of evolution. Fischer's mathematical tools initiated the foundations for the subject of bio-statistics very much used in medical (biology and psychology) as well as other related fields. Haldane's idea of 'primordial soup' for the evolution of life formed the basis for understanding the possible origin of life from inorganic molecules. Considering the impact of mathematics on research in other areas it is worth noting the use of the well-known Fourier Transforms in X-ray crystallography, astronomy, and almost every case of the stability analysis of various physical configurations. The FFT (Fast-Fourier transform) technique is used for fault analysis, quality control, and machine monitoring. It seems to be a central feature in the Wi-Fi technology which nowadays is a must for every mobile phone and computer user. Another topic of interest that life scientists are interested in is theoretical ecology which offers tools to analyze biodiversity and diseases using the mathematical techniques developed by the theoretical physicist Robert May (Shaver, 2018).

As said earlier while discussing the language of science, communicating science depends upon two main components- theories and models. Theories, sometimes considered as stories are generally a summary of the behavior of natural systems and allow one to generalize the possible changes that could occur in different circumstances. Models are considered as an elaboration of certain particular aspects of the theories developed to elucidate specific issues addressed in the theories. In biological systems it is believed that any organism's characteristics (morphology, physiology, size, behavior, etc.) are governed by natural selection and the models used for analysis are mathematical, analyzing the data available using methods of calculus, differential equations, matrices, and such simple tools. Using computational soft wares like Mathematica, Mathlab might ease the process of calculations but knowing the mathematical principles behind such soft wares would increase the understanding to a much greater extent.

Is Mathematics Difficult to Learn?

It is to be appreciated that mathematics is a language, with its own grammar and nuances based on logic and ingenuity. Once understood, it works well in developing a special and clearer view of the world. Mathematics is not just a collection of theorems, statements, and proofs to be learned and memorized for reproduction in examinations/interviews. Mathematics is as much a lively subject as other sciences which additionally has a body and soul that can stimulate every intellect and provide new trajectories for learning the truth and beauty of Nature. As may be experienced mathematical logic helps in building theories in various other branches of science as well as in other human endeavours. Still very often one hears complaints that mathematics is very abstract. Yes, several concepts appear difficult to follow when confronted with for the first time but once illustrated with examples it gets possible to bring them to the concrete domain by looking at their applicability. Unless one is into research, the mathematics one normally comes across is far from being abstract. The operations involved could be a little difficult to follow until one starts using them for solving a problem. Before solving a problem one should first look at the quantified data and express it in terms of different variables. Once such identification is made, one can look for the number of unknowns and possible relations among the different unknowns and the provided data. Sometimes there may be a mismatch between the number of unknowns and the number of equations that could be set up among the variables in which case there may not exist a solution to the problem. If the unknowns are more than the number of equations the system is called underdetermined and could come under a special type called Diphontaine equations which sometimes are solvable with ingenuity. If the number of unknowns are less than the number of equations, the system is called over-determined and there could be possibility of having more than one solution.

More often than not, many theoretical explanations of physical phenomena involve differential equations that generally show the evolution of the system either in space or time, and to solve them one makes use of what are called initial conditions or the boundary conditions. Equations themselves are termed as 'equations of motion. Well-known examples are in Newtonian mechanics. After the computer revolution with high-speed computers and several languages and codes, many soft wares exist that primarily aid in solving such problems. But it is always useful to try and work out an analytical solution before adopting numerical procedures to solve the system of equations. This is so because of the reason analytical solutions offer deeper understanding of the system under consideration as compared to numerical solutions. Of course, such a situation may not be possible in the case of highly non-linear systems which due to their self-interaction may not be amenable to analytical solutions. In fact now-a-days, research in many areas are governed by non-linear equations which may tend to be chaotic and require numerical simulation for even a basic understanding. The theory of weather prediction and climate studies are particular examples of such systems which have sets of coupled non-linear second order partial differential equations with parameters that are continuously self-interacting and changing. Another example of similar difficulty is faced in the context of high energy astrophysics in the context of structure and stability of extremely compact objects that has to deal with analysis of general relativistic three dimensional hydrodynamics governed by coupled non-linear partial differential equations. However these are difficulties faced at the research level and not in following basic sciences. Leaving aside these difficulties involving Mathematics, it can be seen as a way of thinking that develops through the association of numbers, shapes, and simple experiences of ordinary day-to-day living.

Can one see a role for Mathematics in daily life?

Numbers do play a prominent role in all transactions that happen every day in anyone's life. All measurements are in terms of numbers which are used to compare and contrast the sizes, distances, and valuations. Starting with making of morning tea, one needs to add measured quantities of tea, milk, and sugar to a given volume of boiling water. Each one of the ingredients has to be in the correct proportion to provide a good cup of tea. Similarly while cooking, all the recipes prescribe definite amounts in terms of the quantity of each ingredient and the best results appear only when proportions of different ingredients are in the correct ratio added at appropriate stages of cooking. Good cooks learn this with experience.

One of the interesting facts one comes across is the capacity of mental arithmetic of the fruit and vegetable vendors, who calculate the total amount of money to be charged for all the items bought by the customer without recourse to pen and paper. This involves the actual operations of multiplication, addition (also subtraction appropriately, when the customer hassles), and the chances are that the vendor may not have gone to any school but still does his arithmetic precisely.

Life depends upon the occurrence of day, night, week, month and so on that follow from celestial mathematics involving, ideas and formulae from arithmetic, geometry, trigonometry, and calculus of the spin and orbital motion of the Earth around the sun. In today's world of satellite and space technology, mathematics is a basic necessity to compute the forces that a space probe would encounter. This is required for calculating the exact orbital parameters of the satellite to ensure course corrections in near real-time and keep the satellite in orbit and be observable from ground stations at all times. Any miscalculation in this regard may turn out to be very expensive and even endanger the lives of astronauts.

Physical fitness also requires definitive numbers, to track the exercise regime, the distance one runs, the pulse rate, the body weight, and the recovery time. The trainer keeps a record of these to develop healthy exercise regimes required for fitness training. Another interesting application is the determination of heights and distances used by engineers and mountain climbers. The fact that one can determine the height of a hilltop by using a meter-long wooden stick is a good example of the use of trigonometry and Euclidean geometry. It is quite well known that the Egyptians used the 'Pythagoras theorem' and trigonometry to design and calculate the dimensions and angles for the construction of the tombs of emperors in pyramids. Arranging one's living room also requires the knowledge of shapes and sizes of the furniture and their adjustments to the dimensions and the design of the room. When initiated, one can start looking for the inherent mathematics in day-to-day functions and understand its pervasive nature.

What Is the Best Way to Learn Mathematics?

It is by doing and practicing. As it is a symbolic language one needs to be very specific in associating facts with figures and understand the difference between the data given and the logical deductions derived from it. It is always necessary to write down the statements and identify the known and the unknowns which help in setting up the equations that need to be solved. Students should be encouraged to try and link their informal knowledge and experience to mathematical abstractions and this can and should be followed through paper or wooden models representing puzzles. It is often said by researchers that one learns mathematics better when one enjoys doing it. This enjoyment can come particularly while solving problems in Euclidean geometry and Trigonometry which indeed provided the basic foundations for the development of mathematics apart from logic and analysis.

A very interesting feature of numbers that gets reflected in Nature is the Fibonacci sequence, which goes as [0, 1, 1, 2, 3, 5, 8, 13, 21, 34, ...] where every subsequent number is the sum of the previous two numbers. Isn't it interesting to realize that the plant 'cactus' grows its leaves in this sequence? One can also find the numbers of the Fibonacci sequence in the spirals formed by individual flowers in the composite inflorescences of daisies, sunflowers, cauliflowers, and broccoli. Another interesting example in this context seems to be the reproduction of rabbits in their natural habitat.

This sequence can be expressed in the form,

$$F(n) = F(n-1) + F(n-2), \text{with } F(0) = 0 \text{ and } F(1) = 1.$$

It should be very interesting to find out that the limit of this sequence as n tends to infinity is the well-known 'Golden ratio', generally denoted by the Greek letter, $\varphi = 1.6180339887\ldots\ldots$ This has numerous applications and in particular architects and engineers use this ratio while designing structures as it gives aesthetically appealing designs. Any rectangle with the sides having this ratio (length/width = φ) is called a *golden rectangle*.

Apart from all these, mathematics can be fun too, if only, one learns to appreciate the logic and diversity that one has to go through while solving puzzles and brain sharpeners!

Fun with Mathematics

It is admirable that in recent times the daily newspapers carry puzzles like sudoku which has caught the imagination of young and old. There is no better way of appreciating the number system and its variety than to design and solve such puzzles. In this context of properties of a sequence of numbers an anecdote about one of the greatest mathematicians of the last millennium, Karl Friedrich Gauss of Germany is worth a mention. Gauss was about six years old and in primary school. Once to keep the children from making noise in the class the teacher asked them to add numbers from 1 to 100. No doubt, the class became quiet and all the children except Gauss were busy adding. When the teacher chastised him for not working, Karl said "Mam I have already completed, the answer is 5050". The surprised teacher asked him how he did without writing, Karl replied—Mam, 1 + 100 = 101, 2 + 99 = 101, 3 + 98 = 101, ... 50 + 51 = 101, So the answer is 50 times101 which is 5050. The teacher had no response except to realize the budding of a great mathematician. No doubt one knows now that the sum of n terms of an arithmetic progression starting with 1 and with common difference 1 is [n (n + 1)/2].

A key message in this story is the methodology in approach to a problem: Gauss had noticed the repetitive nature of adding successive numbers from the two ends of a sequence, as the ascending and descending sequence should add up to the same total. Such stories when told to children before starting the arithmetical operations may induce them to think originally instead of just remembering a given formula.

Another story of similar nature but involving geometry tells about a boy (less than 5 years) who was seen by the mother as cutting paper into different pieces. He was scolded for waste of paper and time. When his father heard of it, he was curious to know and looked at the shapes of the pieces which were all triangular. The boy was simply putting the three edges of the triangle together by folding and had found that irrespective of the size and shape of the triangle, the sum of the three angles was always 180°. This became a famous theorem in Euclidean geometry and the boy grew up to become the most celebrated mathematician Euler.

Let me quote a few other stories as puzzles that should create wonder and enjoyment in solving.

1. An anecdote that catches attention concerns Ramanujan and Hardy, both famous mathematicians. When Ramanujan was sick and in the hospital in England, Professor Hardy visited him, and Ramanujan asked him the taxi number in which he came. Hardy replied, oh it is a boring number, 1729. Ramanujan immediately retorted, oh no! it is a very interesting number, being the lowest number that is expressible as a sum of two cubes in two different ways. **Can you find the numbers**?

2. Diophantus was a well-known Greek mathematician during the first century AD and his major contribution was in number theory. Today we know him only through the famous Diophantaine equations that are an indeterminate set of equations (equations set with the number of equations less than the number of unknowns) which have only integral solutions. An interesting puzzle found on his tomb tells his story with the solution showing his age. "Diophantus, Ah, what this tomb holds a marvel! The tomb tells scientifically the measure of his life, a sixth portion of which was a beautiful childhood. When a twelfth was added, his cheeks acquired a beard. A seventh part he spent in childless wedlock. Five years then passed and he rejoiced in the birth of a son when Fate measured out a joyous and radiant life on this earth only half of that of his father. After consoling his grief by the science of numbers for 4 years he reached the end of his life. How many years did Diphontaine live before death overtook him? **Can you find out his age**? (Perlmann, 1942).

3. A shopkeeper had sold a certain quantity of cloth at Rs 49.36 per meter but while entering in the register, he did neither write the length sold legibly (except that there was no fraction) nor the amount received, with only the last three digits being clear Rs xxx7.28, But a smart chartered accountant found him out. **How**? (Perlmann, 1942).

4. This is about a King and his Vazir (a minister). King had three children and he had decreed that after his death, all the property to be divided among his three children such that the first one gets half, the second one, one third (1/3), and the last one only one-ninth (1/9). When the king died the properties were distributed as per the king's wish. The problem came with elephants as there were only seventeen (17) of them. Everyone was arguing and blaming the king, but the Vazir had an answer. **Can you find it**?

The solutions to these problems are provided in the appendix, but it will be certainly rewarding to try and solve them and realize the beauty of mathematics, which is known among researchers as the 'Queen' of sciences.

Before ending the section, it would be useful to refer again to Thomas Kuhn's (Kuhn, 1962), 'structure of scientific revolution', in which he says that doing normal science is like solving puzzles.

"To a scientist the results gained in normal research are significant because they add to the scope and precision with which a given paradigm can be applied". Still, no one devotes years of research just to developing a new instrument (like a

spectrometer) or to prove that Pythagoras was right. Such investigations are often disliked by scientists because they are mostly repetitions of procedures already carried out before. "Challenging scientific research to a conclusion, achieving the anticipated in a new way requires all sorts of the solution of complex, instrumental, conceptual and mathematical puzzles. Person who succeeds proves himself/herself as an expert puzzle-solver and the challenge of the puzzle is an important part of what usually drives him/her on". Almost all fundamental discoveries come under such a classification of scientific research. It may not be wrong to consider Mathematics as one of the fundamental requirements for good scientific research in any of the practiced disciplines. A very useful example in this context is the discovery of planet Neptune in 1847 by Galle, and Adams independently but guided by the precise mathematical calculations of LeVerrier. He was concerned about the irregularity noticed in the orbit of the planet Uranus and attributed it to the perturbations due to the presence of a nearby mass as required by Newtonian calculations and the heliocentric model for the solar system. It is said that while Galle of the Berlin observatory was encouraged in his search by the calculations of LeVerrier, Adams in England had worked on the same problem a few months earlier and had worked out the possible path of the perturbing body. LeVerrier had succeeded in calculating the position (distance from Uranus) and the mass of the body and even had called it a new planet. While the credit for the discovery goes to both Adams and LeVerrier, the main message is the importance of the significance of patient mathematical analysis that led them to it.

Another discovery of similar nature was that of Halley 's Comet. Apparently Newton's discovery of the law of gravity was published only because of Halley's support, who extending the ideas of Kepler regarding the orbits of planets, had calculated the cometary orbits. He must have conceived the idea of orbits of objects under a central force are conic sections and subsequently worked out the long period elliptical orbits of comets. On that basis he searched the earlier records and found similar features in a comet that had appeared in 1531, 1607 and 1682, that encouraged him to predict its reappearance in 1758 (Howell, 2017). Though he did not live to see the return of the comet, his calculations about the orbits confirmed Newtonian mechanics (as Kepler's laws had done), and the comet was named after Halley, yet another success of analysis and prediction. The story of these discoveries is indeed a confirmation of an existing paradigm-Newtonian theory of gravity and the orbital mechanics derived from it. In the modern era with the availability of very high-speed computational help solving such problems with minimal data but good mathematics is much easier.

In summary one can say that there are two types of researchers in mathematics. Those who do mathematics for its own sake like Hardy and Ramanujan, and those who work it out to check or verify some principle or discrepancy that could sometime give a new result. Either of these approaches bring in a lot of satisfaction and continue the cause of science methodically.

References

BSS. (2017). *Breakthrough science society.* A brief history of Science. Society.

Courant, & Robbins. (1941). *What is mathematics?* Oxford University Press.

Howell, E. (2017). https://www.space.com/19878-halleys-comet.html

https://en.wikipedia.org/wiki/Babylonian_mathematics

Kuhn, T. (1962). *The structure of scientific revolutions.* University of Chicago Press.

Perlmann, Y. (1942). *Mathematics can be fun.* Mir Publishers, Moscow.

Prasanna, A. R. (2017). *Gravitation.* CRC press U.S.A, for Taylor and Francis group U.K.

Shaver, P. (2018). *The rise of science.* Springer.

Teicher, A. (2014). Mendel's use of mathematical modelling: Ratios, predictions and the appeal to tradition. *History and Philosophy of the Life Sciences, 36*(2), 187–208. JSTOR, Accessed February 16, 2021, from www.jstor.org/stable/44471280

Chapter 6
An Integrated Approach to Communicating and Practicing Science

Introduction

In the prevalent educational system, Science teaching has been compartmentalized into subdivisions, which often leads to even interrelated topics being treated separately. Such subdivisions, without consideration to overlapping continuity of concepts, are against the spirit of science and fail to relate science as an integrated study of different aspects of any phenomena thematically. If one observes Nature and the events that occur through experiences and tries to understand the underlying theme, then one can learn the whole phenomenon in an integrated form. This avoids just remembering the words or equations that describe a phenomenon and keeps the basic theme which unifies the different aspects of the occurrence intact. Taking the example of the theme 'motion', one can combine learning all the different aspects that govern motion through different examples that one sees in nature and experiences. Linking different aspects like walking, crawling, running, flying to the anatomical features of different species, one can make learning both the physical and biological aspects of 'motion' together. Continuing the trend one can relate motion with the concepts of 'force' and 'work', which then links on to 'energy' all of which can be made illustrative than mere 'word definitions'. With the newly acquired awareness about the importance of understanding the environment and ecology, almost all the basic features of physio-chemical effects of matter and their link to life forms have many examples that one can use to make the student understand the differences between the science of the living (organic) and of the non-living (inorganic) in an appropriate manner.

Modern science taught in schools and colleges, divide and subdivide the themes into compartments and sub-compartments like mathematics, pure and applied physics, chemistry, biology, geology, astronomy, etc. This is unfortunate but perhaps unavoidable due to the vast amount of information a learner is expected to assimilate and then reproduce verbatim in the examination for evaluation. But has the student learned science? Did the school lessons convey the meaning and intricacies of science as they should have been? Is it not necessary to impart the knowledge of Nature as a whole with its interconnectedness and not in parts? This does not mean that one should burden the young minds with excessive information memorizing which may lead to remembering the words but receiving no scientific knowledge. As long as one treats any subject only for its information content this predicament is unavoidable. One needs to convey the concepts of science in as united a manner as possible. This is doable at least for some of the basic aspects of nature such as matter, energy, work, force, and such concepts one meets, and experiences in everyday life. The suggestion here is to gather/provide sufficient information about a phenomenon and its associated processes to appreciate the interconnectedness of the processes with experiences that help one to get curious enough to ask questions for a better understanding.

Biographies of the leaders in science ranging from Galileo to Einstein or Darwin to Khorana, indicate that the correct approach in science is to look at any phenomenon in its entirety. Aristotle, considered the pioneer of scientific research, though is acclaimed for his contributions to astronomy, as it is known today, made his most lasting scientific contributions in biology and zoology. He lead a large team of investigators, who travelled throughout Greece and Asia collecting specimens of living creatures from sea and land and studying their behavior methodically (Cane, 1961). Of their several conclusions, the most important was the recognition that there is an order in the world—the ladder of nature where the living beings could be classified by their complexity and are fitted to the conditions of their lives and have a functional perfection—a revelation that helped him propose and extend the geocentric system of the Universe, a theory he had learned apriori from the works of Plato's other disciples.

What aspects of science need to be taught at the primary/secondary levels of education? Just enough to appreciate, understand, and protect the nature around and the environment one is living in. To start with, one needs to express the lessons thematically and not as mere topics! The term thematically implies the aspects related to the daily experiences of individuals and/or groups which can be translated into words and explanations provided with the help of the varying physical, chemical or biological examples when probed. The experiences one refers to are in general responses to one's actions. In the context of understanding animal behavior, it is important to observe their behavior in their natural surroundings which will keep them better responsive than putting them in constrained laboratory conditions. Luckily the new discipline of 'behavioral science' seems to serve this purpose by appropriately choosing themes to categorize the animal behaviors. Similarly in various other aspects of understanding nature, the inter-relationships among events can provide recognizable themes which can be and should be analyzed. In learning science, it is not necessary to remember/memorize, but to visualize and narrate as well as connect the description for any successive event that could have a similar cause-effect relation. Once a theme is thought of then analyzing it into different sub-themes can lead to the required procedures like just observing samples or devising experiments that can clarify the cause-effect relation.

The simplest of the notions to appreciate and understand is the concept of Motion. From the time a child starts crawling, followed by walking, the purpose for the action is either to reach for something or get away from someone/some happening. Once the purpose is well defined the action is automatic. To understand the dynamics of motion, the concepts of space, displacement, and time are needed. Linked to this is the concept of speed and the related notions of slow and fast. Speed has only magnitude. By introducing the concept of direction and associating speed and direction, the notion of velocity and the concept of vectors can be learned. While introducing these concepts quantitatively it may be demonstrated how the physical quantities are expressed in terms of mathematical entities like the numbers, lines, curves, graphs, etc. as well as the notion of scalars and vectors. The concepts of uniform motion along straight lines and of accelerated motion along curves is only a logical next step (the path of a projectile is a parabola; it is a common experience that when one throws a stone randomly the path it makes in the air before reaching the ground is a parabolic arc). Analyzing these aspects in a combined form will include aspects of physics (displacement, speed, velocity), mathematics (geometry, trigo-nometry, algebra) with concepts learned than memorized. Continuing the same theme, it should not be difficult to understand the anatomical and physiological differences between animals and birds in respect of motion. Land animals have limbs to walk or crawl, sea creatures have fins to swim and birds have wings to fly. In the context of 'motion', questions like-why would one move fast? Or when would one move fast? And how would one move fast? can weave the entire story of motion with its physical, mathematical, and biological features and can be told holistically to propel young minds to more questions that may lead to new insights. The anatomical differences among the various species also must have evolved with life and the environment. Having discussions on such topics is the best way to create interest and curiosity among youngsters which can go a long way in making them realize, what they might want to pursue for higher studies.

Closely related to the concept of motion are the concepts of force, work, and energy. It would indeed be easy to define these terms in simple words and make children internalize them. Further, making these concepts activity-oriented through simple and practical examples that involve students themselves can make the concepts visible. Most of these scientific terms do appear self-explanatory and illustrating them with action leads to asking why and how. Communication of scientific knowledge is most effective when it can be contextualized to events and processes around the learner.

The ideal way for learning science is through examples from different sub-areas but related to a single concept. For illustration consider the theme structure and constituents of matter. As is known matter comes in different forms i.e. solids, liquids, and gases. (Of course, there is a fourth state—plasma but this is not introduced at the elementary stages of education). The transition between different phases of matter happens through the exchange of energy as well illustrated by ice, water, and steam. Heat energy is the agency that helps in the transition from one form to the other. The arrangement or freedom of movement of molecules in a given state is related to the amount of heat energy appearing as the energy of motion (kinetic energy) of molecules. The water molecule is made up of atoms of hydrogen and oxygen. As one looks at the structure of these atoms individually, one sees a difference in the number of their constituents and structure. Taking the theme of atoms and molecules, one can try and unify the introduction to physics, chemistry, and biology of matter and energy, inorganic and organic, the non-living and the living! Once such a unified approach gets clear the learner is ready to ponder on deeper questions and the aura of scientific investigation will take its roots. As another example for 'thematic presentation', one can choose the topics 'states of matter', 'elements, compounds and mixtures', and 'separation of substances' which are classified as different chapters in some school-level science textbooks. Looking closely at these topics one can see that linking these lessons with appropriate examples would make it interesting to learn the basic science of the three aspects together and identify the structure of matter, its constituents, and their energy of binding. This example can very well illustrate the differences between physical and chemical bonding. Textbooks have their lessons on water and air which are indeed the best examples for compounds and mixtures, and form basic entities for life on earth. The concepts associated with particulate matter (atoms and molecules) are basic to both living and the non-living. The basic entity of life forms—the cells also are constructs of more fundamental entities like amino acids and protein molecules which have their structure from atoms of different elements.

A very revealing theme, for understanding the most important energy in our lives, 'Light', is the topic 'Eye-Colour and Vision'. This as a theme can incorporate the physics of light and the perception of color requires learning the role of 'rods and cones' the most important physiological features of the eyes along with the lens and the retina. While discussing the anatomical and physiological aspects of eye the organ, understanding the visual perception of an object through image formation can be discussed using the knowledge of propagation of light and image formation which is related to the phenomenon of refraction by the lens. Many school textbooks do have material that content-wise is very interesting and factual but the presentations of the facts do not seem to excite the curiosity or bring in the connected aspects of different sub-topics in a unified manner. Overemphasis on marks makes students learn the facts by rote, rather than visualize, appreciate and understand the science behind them. While on this topic, it is important to review the attempts going a step further, to the new approach suggested—STEM education which recommends the pursuit of Science, Technology, Engineering, and Mathematics in a closely interactive way (DER18). As pointed out although the idea of integrated STEM claims to offer students an opportunity to experience learning in a real-world multidisciplinary context, unfortunately does not reflect the natural interconnectedness between disciplines which could have consequences for student interest, knowledge, and performance (Moore et al., 2014). It is mainly the lack of background and knowledge of science teachers in the engineering discipline that causes the rift as pointed out by Cunningham and Carlsen (2014) as also perhaps due to the very limited mathematics knowledge of engineers. As mentioned some researchers have reservations in examining the role of technology in STEM due to the complex reality of defining technology in education (Herschbach, 2011) and none of the proposed models seem to provide the learning facilitators details regarding instructional strategies for the classroom deployment of such a system. This practical difficulty can perhaps be circumvented if the proposal is taken in steps. It would be far easier and useful if only the science and mathematics are integrated through the thematic presentation as mentioned in the beginning in classroom teaching up to the higher secondary and bring in technology and engineering at the next level. It would be certainly worthwhile to keep giving examples of the basic scientific principles that guided technological applications during the science courses preferably with examples of innovations inspired by pure scientific understanding. Such an approach would help create learners' interest in all aspects of STEM education.

Even from the early times, scientists who pondered on questions regarding Nature did so in all its manifestations together and not confined to just one. It is true that in the present century, the information concerning every field is so vast that super-specialization often becomes necessary. But that is needed only at the research level. As far as basic science learning it should still be possible to look at events and phenomena in a closely related manner. It is good to know that of late there is an increasing effort to teach science in an interdisciplinary manner. As an example, one could consider courses related to environmental studies, where one should be able to correlate the importance of ecosystems in the various habitats and their relation to climate and other geographical and biological features. The physical, chemical, and biological features of different species can be illustrative examples for conveying ideas related to life evolution and diversity. Specific inputs from mathematical, physical (including geological), chemical, and biological aspects of our life and surroundings need to be kept in mind to encourage questions on aspects that excite students and guide them to learn the associated facts and not just believe in what was told or what is written in a book. The educational system must allow enough time in the school curricula for increased discussion among students and teachers. This would need changes in the current examination/evaluation practices, which can ensure that the final results would be very beneficial to all concerned. These changes would be possible only when the boards and committees that recommend the framework of syllabi for different levels of education have regular teachers who face the classroom with students as members and not just administrative officials and subject experts who may not have had the facility to interact with students. In this context I am reminded of a very funny but unfortunately true opinion, a senior colleague had mentioned. In the existing system teachers from the primary level to college level are always under the tension that they have to 'cover the syllabus' in the restricted periods as otherwise there can be a black mark. Is it not the duty of teachers to 'uncover the topics' in the syllabus and make students learn?

This topic of the integrated approach to science learning is perhaps a difficult and intricate effort to bring in its implications fully to practice. Numerous discussions must have happened at various levels of teachers and experts who would have agreed on the necessity of such an approach. But putting it into practice is a cooperative effort for practicing teachers (learning facilitators) and administrative authorities in charge. It would be useful to have teacher groups set up and have workshops to discuss and evolve examples of themes that require inputs from different disciplines.

Such interactive discussions would surely help in creating material for communicating science at different levels optimally. The difficulty may appear mainly in identifying the themes that are expressible in terms of simple, age-appropriate concepts that students would follow. This calls in for facilitators who are themselves good and well informed in different aspects of science and at the same time have motivation for research. This would require minimising the information content given usually as advanced material. If the students understand the themes with appropriate examples then they get motivated to look for needed information from various sources that are easily available on the internet and associated search engines. At this point proper guidance from the teacher regarding 'on line search' is needed, without which the students may be led astray.

It may be useful to point out certain important aspects highly recommended by a few educators in this connection. To start with a few quotes as pointed out in an article by Nitu Kaur (Kaur, 2016, 2019) "The very notion of 'integration' incorporates the idea of unity between forms of knowledge and the respective disciplines (Pring, 1973)". R. W. Emerson (Emerson, 2014), American philosopher says "The astronomer discovers that geometry, a pure abstraction of the human mind, is the measure of planetary motion. The chemist finds proportions and intelligible method throughout matter; and science is nothing but the finding of analogy, identity, in the most remote parts". It has been often pointed out (NCF, 2000) that 'science is the creative response to the curiosity and capacity to wonder present amongst every human being. Learning of science in schools augments the spirit of inquiry, creativity, and objectivity along with aesthetic sensibility. Patterns and relations, make and use new tools to interact with nature, and build conceptual models to understand the world'. This human endeavor has led to modern science. What could be detrimental is the unfortunate feeling that some students may get by the time they are in the higher secondary that 'school has little connection to their current lives and even less to offer in preparing them for the future' (Clayton et al., 2010). It is thus very important to make understandable and relevant connections between the topics they learn and the life they are going to face. Dr. Kaur concludes her essay by saying, "Fragmented approach is having a narrow approach for learning and it does not ensure psychologically sound learning whereas integrated approach provides a larger canvas for the learner's discourse and interaction with strong inter-linkages in between interdisciplinary themes and concepts". She has also provided an example of a biology lesson on 'photosynthesis' that can be linked to topics in physics and chemistry which convey the need to integrate the science topics thematically and enthuse the learners to making a possible integrated approach to science learning. It is necessary to point out that the attempts to integrate curriculum is not new as one sees from the article by R Pring (1971) who said that "the unity of knowledge or a single view of the world and of life can be reflected adequately only in an integrated curriculum. Though there are wide differences in approaches, all have in common a disapproval of fragmentation of the curriculum (that typifies traditional school) that leads to subject barriers and the compartmentalisation or pigeonholing of knowledge, accompanying specialisation and frequent irrelevance to real problems. This brings in a need for a critical appraisal of different proposals and a closer examination of how the word integration operates in discussions on educational systems".

A recent article by S. Dattagupta (2020) in the same context tries to bring out certain lacunae existing in the system of higher secondary education which needs to be looked into and adopted. As he points out separating biology and mathematics at the school level could get detrimental for students who want to go in for interdisciplinary subjects like molecular biophysics, bio-mathematics, chemical kinetics, and similar subjects which are very important for understanding new developments in life sciences. Similarly, the advanced developments in medical technology need an understanding of computer technology and various languages for interpreting the results of investigations. Most of the time there are indeed ready-made soft-wares available which makes it easier, but a basic knowledge of the related discipline would make one self-sufficient in understanding the cause and effect relationship more clearly.

Modern biology and medical courses require a good deal of understanding in the linked basic areas of physics, chemistry, and mathematics as hinted in the previous section. In the title, it was expressed that this approach is needed while practicing science too. One may ask, won't that distract the investigator to be looking at more than one discipline that may dilute the concentration? Of course not. If one looks at the history of scientific discoveries, most of the time it is with the application of combined knowledge from different disciplines that new findings were made. One of the best examples of such a finding is the structure of DNA by Crick, Watson, Wilkins, and Rosalind Franklin, and its applications. The stimulus for the investigations initiated by the physicist Francis Crick came from the book by Erwin Schrodinger on 'What is Life' where he has summarized his earlier lectures on the question of how the basic physics and chemistry within a cell could explain the secret of life. The final answers indeed came with contributions from Physics (X-ray crystallography), Chemistry (the molecular structure), and Mathematics (the topology).

Research in oceanography and its impact on climate and weather is another important topic where one needs data and interpretations from the physical, chemical, geological, and mathematical features taking into account the varied influx of pollutants from agricultural, industrial outflows, and fossil burning, along with other possible natural causes. The study and research in this area can be very useful particularly for communities living in the coastal regions and the fishing community to a very large extent. As studies in the last few decades have shown most of the natural systems are self-interacting and thus exhibit a non-linear behavior that can become chaotic. Over the last thirty to forty years mathematical models of non-linear systems have been a flourishing subject and getting to understand and contribute to its development does require a concerted effort of following the associated science in as much an integrated way as possible.

Though it is a very difficult subject to bring in for a general reader, it is important to note that scientists have been thinking of understanding nature in all its perspectives with barriers removed. Such a venture is to yield results of significance, future generations must be initiated to look for explanations of observed or inferred events, considering all the different aspects of science in an as interactive and unified manner as possible. Unification of ideas on a given theme is a very important aspect of learning science.

References

Cane, P. (1961). *Giants of science*. Pyramid books, Grosset and Dunlap, Inc.

Clayton, M., Hagan, J., Ho, P. S., & Hudis, P. M. (2010). *Designing multidisciplinary integrated curriculum units*. ConnectEd: The California Center for College and Career.

Cunningham, C. M., & Carlsen, W. S. (2014). Teaching engineering practices. *Journal of Science Teacher Education, 25*(2), 197–210.

Dattagupta, S. (2020). *Current Science, 118*(9).

DP17. https://baranlab.org/wp-content/uploads/2017/06/DNA-Chemistry

DS09. http://www.math.pitt.edu/~swigon/Papers/S_ IMA.pdf

Emerson, R. W. (2014) *Essays*. Xist Publishing.

Herschbach, D. R. (2011). The STEM initiative: Constraints and challenges. *Journal of STEM Teacher Education, 48*(1), 96–112.

Kaur, N. (2016). Learning science is all about getting the concepts right. *Mizoram Educational Journal, 2*(1), 57–68.

Kaur. (2019). JETIR March 2019, Vol. 6, Issue 3.

Moore, T. J., Stohlmann, M. S., Wang, H.-H., Tank, K. M., Glancy, A., & Roehrig, G. H. (2014). Implementation and integration of engineering in K-12 STEM education. In J. Strobel, S. Purzer, & M. Cardella (Eds.), *Engineering in precollege settings: Research into practice* (pp. 35–59). Purdue University Press.

NCF. (2000). National Council of Educational Research and Training. National Curriculum Framework.

Pring, R. (1971). https://doi.org/10.1111/j.1467-9752.1971.tb00455.x

Pring, R. (1973). Curriculum integration. In R. S. Peters (Ed.), *The philosophy of education* (pp. 123–149). Oxford University Press.

Chapter 7
Experience and Experimenting in the Development of Sciences

Introduction

Science is based on a logical interpretation of results derived from experience, experimentation, and observation. Children start learning about things by wanting, touching, and experiencing the result. These experiences make them curious and initiate a spirit of inquiry. Thus already at birth and childhood children have the motivation to learn. Adolescence is the time when children not only want just experience but also try and experiment to know about everything new. Unfortunately currently practiced system of education often curbs this 'wanting to know curiosity' of children and rush them through a fixed regime of 'finishing the responsibility of covering the syllabi' for a predesigned curriculum which does not give enough time either to the teacher or to the taught to ponder and discuss. The very expression of 'covering' the syllabi is self-defeating as the idea of teaching should be to 'uncover the facts' discussed in any lesson. Indeed, the educational institutions at the primary level may not have adequate facilities or funds to perform laboratory experiments. But one should not forget that 'Nature' provides several situations in daily life where basic aspects of science can be conveyed to illustrate the principles underlying one's actions/reactions. The practice of science and technology has witnessed from the beginning illustrious examples of stalwarts like Galileo, Newton, Marie Curie, Fleming, Edison, etc., who all devoted their lives just to understand and repeat efforts to confirm what they had discovered, using the methodology of experimenting and verifying the truths about nature.

Science and technology derived from it are an integral part of our lives. Be it electricity or the thermometer, day-to-day living or travel aids or mobile phones and television, all the facilities we are enjoying have come from the efforts and hard work of men and women who understood and applied science to invent gadgets. Still, there are situations where a quest for a logical explanation of events is needed as such explanations might lead to deducing reasons and inferences for other similar events. Science is a bootstrap process of hypothesis, experiment, observation,

© The Author(s), under exclusive license to Springer Nature Switzerland AG 2022
A. R. Prasanna, *How to Learn and Practice Science*,
https://doi.org/10.1007/978-3-031-14514-8_7

inference and the pursuit never ends. It is necessary to keep in mind that there have been times when some new experiments have led the investigator to the unexpected discovery of unforeseen consequences which is not desirable and on the contrary got harmful in the wrong hands. This aspect of scientific research needs to be carefully kept in mind before new projects are supported and followed.

As science is a community-driven effort, often the experience that one talks about for explaining an occurrence need not be one's own but learned from someone else. This is true for scientific explanations and one does not want to accept all one hears without verification. Failure to practice such a procedure often leads to superstitions and gets implanted as blind faith. This is particularly true when people follow rituals without asking for reasons for doing what was told. A scientist should accept any statement only if it is verifiable by him/her through experimentation or reasoning through deductive logic matching his/her understanding and experience.

Some facts are learned as a part of growing up. These include feelings of thirst, hunger, hot, cold, bright, dark, loud, soft, etc., which are individually experienced. Learning concepts of science from experience is termed non-formal education while the concepts learned from textbook/teachers is referred to as formal education. Such a distinction has increased the gap between understanding and learning (by rot), as children are often directed to learn facts as given in a textbook and reproduce them without questioning. Many times this could be outside their experience. It is always easier to learn when the phenomena can be contextualized with real and local examples.

Many of the scientific principles indeed learned during early education may not be amenable for simple experiments. This is further made difficult due to the absence of laboratory facilities in all schools for doing experiments. Under such situations, the onus is on teachers to try and develop newer strategies that could make the children visualize. The ingenuity of the teacher comes in handy only when she/he can use the common day-to-day experiences at home or in the outside world to introduce scientific concepts. To illustrate, a simple example would be a lesson on the transport of heat. Heat is a form of energy and there is always the exchange of energy by a body with its surroundings. The easiest way to communicate this concept is to ask what happens when the child touches hot or cold water. This experience is a normal reaction of feeling hot or cold that explains the transfer of heat energy from outside to the body or from the body to outside depending upon the feeling. The transfer when it happens by direct touch is explained as conduction. This example can be fortified by showing how heat transfers from the stove to the vessel and from there to the stuff inside, through the transfer of heat energy from molecule to molecule. In this context, the idea of close packing of molecules in solids can be introduced as helping transfer heat faster in solids. The experiment in the textbooks with jar of water and ink demonstrates the transfer of heat by convection, as the distance between hotter molecules gets larger and the space gets filled by the cooler molecules and the hot water rises to the top while the cold water comes down. In the bathroom, one knows the effect of adding cold water to a bucket of hot water. This exchange of energy between molecules continues till all the molecules are of equal energy and no further transfer can occur. Then the liquid is supposed to be in thermal equilibrium and the amount of heat content measured by a thermometer gives the temperature of the liquid.

Such a common experience also plays a fundamental role in defining the weather system as already referred to in an earlier chapter (example of morning breeze and convection in the atmosphere). This theme of heat distribution in different states of the matter elucidates the concepts associated with the thermal energy or kinetic energy of motion and the reason why matter exists in different states at different temperatures.

It will be amusing and instructive to experiment with some ice and a thermometer. Keep the ice in a vessel and insert the thermometer to record the temperature. Just by leaving the two in the open one will find that with time the ice melts into water as it gains heat from the atmosphere through absorption of radiation. One can also notice that the amount of space occupied by the water will be more than the space occupied by the ice showing that water molecules occupy more space as they are moving apart from each other as compared to when they were in the form of ice. Now if you start heating the water after some time it boils and turns into steam and occupies much more space than the water molecules did. Also one can easily see that the thermometer reading slowly rises to show that the molecules have gained extra energy the thermal energy which makes them move apart taking more space.

Apart from the heat, light is the form of energy that one can experience and understand easily. Why is light termed a form of energy? As mentioned earlier, the concept of motion is something one can easily associate with energy. The simplest way to understand this is by the fact that when you put on a light switch in a dark room however big the room is it gets illuminated almost instantly, though the source of light could be at one corner. A better example is to see the transition from night to day when the Sun comes above the horizon, and lights up almost half of the globe at the same time! This is because light travels at a great velocity (300,000 km/s or 186,000 miles/s). Anything with a capacity to travel has energy associated with it. Light energy is the most common form of energy that we depend upon.

As heat can be experienced through touch, light is sensed visually. The experiments that explain the concepts concerning light and its propagation are many. It does not need much effort to explain the shade that comes from shadows. When do we encounter a shadow? The best way to explain is to show why some objects make a shadow while some others do not. This again is linked to the propagation or motion of light. Light can pass through some materials (that are called transparent) and cannot through some others (called opaque). There is a third type of material that permits partial propagation of light and is called 'translucent'. Taking propagation of light as a topic, it is possible to introduce concepts like the nature of light, reflection, refraction, scattering, absorption, re-emission, etc. which include the physical and chemical properties of materials through examples from astronomy, physics, chemistry, and biology (eye, vision, and color). Depending upon the mental maturity of students addressed to, one can even introduce the concept that light is a form of electromagnetic wave thereby setting a background for more advanced concepts of electromagnetism and optics.

A good story to cite in this context is that of A.A. Michelson and his quest to find the speed of light on one hand and his experiments with Morley to find the speed of the Earth in the aether considered as all-pervading. Failure of the Michelson-Morley experiment supported Einstein's assumption for the special theory of relativity by ascertaining the fact that aether does not exist and that the velocity of light is isotropic (same in all directions). Although the nature and speed of light had been a subject of much thought and discussion since ancient times, it was only towards the end of the nineteenth century that a proper understanding was obtained in this regard. The earliest of the ideas about the straight-line path of light was known to the Greeks by about 300 BCE. Ptolemy had compiled a table of measured angles of incidence and reflection for different media while Aristotle speculated that the colors of the rainbow could be due to the reflection of light by water droplets. It was then thought that light travelled with infinite velocity till Galileo, the father of experimental science made the first attempt to measure the speed of light but could only come to the conclusion that light travels much faster than sound with very high speed. During the seventeenth and eighteenth centuries, the opinion was divided and prominent physicists like Kepler, Renè Descartes, Boyle, Hooke all had divided notions about the speed of light. A direct manifestation of the finite speed of light was hinted only after the discovery of the aberration of starlight by J. Bradley (1728) and through the time difference measurement of the successive eclipses of Jupiter's moon Io by Olaf Roemer (1676) indicating that the time difference occurred due to the time lapse in receiving the light from Jupiter at the two different locations in earth's orbit. The difference was attributed to the distance covered by light between the two points in Earth's orbit. He even gave a number for the speed to be about 138,000 miles per sec. Actual measurements were first made by Fizeau (1849) and later by Foucault (1878) and Newcomb in 1881. However, the final value for the velocity of light was due to A. A. Michelson (1852–1931) who used an improved version of the rotating mirror technique earlier used by Foucault that was considered the most accurate measurement. Michelson who was the first American Nobel laureate (1901), used an octagonal mirror and had the light beams travel a much farther distance (500 ft) than the earlier efforts, with the mirror rotating at 128 turns per sec. With this set up after several attempts, he obtained for the velocity of light the mean value of $186,500 \pm 300$ miles/s which indeed is almost the value accepted even today. (Michelson, 1879, Klepner, 2007).

Apart from this, Michelson was also studying deeply the phenomena of interference of light which indeed lead to confirming a revolutionary idea in physics for a completely unknown and unexpected reason. During this era, there was a firm belief among many that the Earth is embedded in an all-pervading medium-the aether, and Michelson's idea was to determine the velocity of the Earth with respect to this medium. His idea was to race two beams of light in this aether wind (as two swimmers in a pond) and compare their travel times which could give data about the velocity of the Earth with respect to the aether medium. Towards this, he constructed an interferometer as shown in the figure where a pulse of light is split into two with one going through whereas the other travels in a perpendicular direction. Both the split beams get reflected by fixed mirrors at some distance and

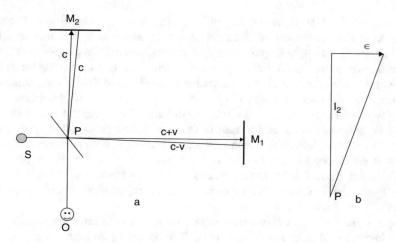

Fig. 7.1 MM interferometer

return to the eyepiece where they interfere and produce an interference pattern (Fig. 7.1).

The rays split at P and go to the mirrors M1(moving along the direction of Earth's motion} and M2 (in the direction perpendicular to the Earth's motion) and they both reflect to P and the eyepiece. According to Newtonian mechanics of velocity addition, the velocity of beam 1 is (c + v) while going out and (c − v) while it comes back, v being the Earth's velocity with respect to the aether. For beam 2 the velocity in both directions is c as it would not be affected by the Earth's motion. By rotating the entire setup by 90°, and repeating the experiment one would expect a slight shift in the interference pattern. The shift in fringes should be proportional to the value (v^2/c^3) which was measurable. Michelson and Morley repeated this experiment several times but found no fringe shift to a very high accuracy. One would consider that the experiment had failed. On the other hand, it had only shown that the velocity additions as required by Newtonian mechanics could not be right. This so considered failed experiment stood as a verification that the velocity of light is independent of the motion of the Earth. This was indeed what Einstein, though much later had assumed for his special theory of Relativity that the velocity of light is the same in all directions irrespective of the motion of the source or the observer. Einstein had removed the absoluteness of time by making it the fourth dimension, and maintained that 'the velocity of light is the same for all uniformly moving observers and in all directions'. Thus the famous negative result of the Michelson-Morley experiment had provided the unique feature of a 'failed experiment revolutionizing modern physics'. (It is worth mentioning here that as claimed, Einstein was not aware of this experiment when he proposed the basic postulates of special relativity!) Further, the constancy of the velocity of light in the vacuum was necessary also for Maxwell's theory of electromagnetism, to describe the existence and propagation of the electromagnetic waves (Prasanna, 2008).

Conversion of energy from one form to another is a daily experience. The most common is the conversion of electrical energy to mechanical (electric fan), heat (electric stove), and light (electric bulb) energy through several gadgets that everyone is familiar with. Such familiarity and experience can be used for the illustration of the basic principles involved and can also be an effective way of communicating science and its role in developing technology. The role of experience and experimenting in creating good science is best illustrated from the life of Galileo. When he was convinced of the truth in the Copernican heliocentric system after seeing through the telescope which was a new invention, he made observations to establish the theory along with discovering the moons of Jupiter. His experiments with balls and the inclined plane are indeed well known to students of science, as the experiment which supports the well-known 'principle of least action' (mentioned earlier).

One of the important discoveries which came out of a large number of trials (experiments) is that of the electric bulb. The names of Humprey Davy (1809), Joesph Snow (1879), and Thomas Alva Edison (1880) are associated with this. The interesting story is ascribed to Davy, who with his team tried numerous materials for making the filament for the electric bulb. When none had worked he seems to have cried in disgust, "Oh! What a shame, I have tried everything except this" and saying so spread out his hands which were full of coal dust. But then immediately he seems to have said, why not try carbon! He then attached a carbon strip between the ends of a wire attached to a battery and found it glowing. However, it was noticed that this was not efficient (did not last more than a few minutes) neither some of the other attempts with platinum filament which was too expensive. Finally, it was left to Edison (1880) to use a carbonated bamboo filament that glowed without burning in an oxygen-free atmosphere for over 600 hours thus making it commercially viable. It was in 1913 that Irving Langmuir discovered the effective use of an inert gas-filled bulb instead of being just an air-free bulb that finally resulted in the incandescent light bulb (endesa17). Later it was found that the element tungsten (with very high melting point and thus lasts longer) was best suited for the filament (1906). Now a days the LEDs have indeed taken over in a big way which has replaced most of the other types both commercially and at homes.

The efforts of Marie Curie and Pierre Curie, as experimenters are unparalleled. Marie earned two Nobel prizes, one for Physics (1903) sharing with her husband Pierre Curie and H Becquerel, for the discovery of radioactivity and later for chemistry (1911) for the discovery of the element Radium. With bare minimum facilities, she worked in a shed, to separate a few grams of radium from tons of pitchblende, the ore of Uranium. This involved continuous boiling of the molten liquid and filtering it again and again for refinement. Apart from identifying Radium, Marie and Pierre had also extracted, Polonium (named after Poland, her country of birth) and a compound of Bismuth. As all these elements are highly radioactive, Marie Curie died of radiation-induced leukemia, and thus sacrificed her life for science and is revered forever.

Another remarkable story is that of Jocelyn Bell of Cambridge, England discoverer of Pulsar (1967) (Hewish et al. 1968), the pulsating radio source. A pulsar is described to be a fast rotating neutron star that emits pulses of radio waves with the most accurate periodicity. Though the emission mechanism is still debated upon, it is theorized (Gold, 1968) that the radiation could be a result of a rotating neutron star, the end product of a dying star having a mass of 1.5 to 4 solar masses. It has an enormous magnetic field (almost 10^{12} Gauss which is more than a trillion times the Earth's magnetic field), produced mainly from the trapping of the magnetic field lines of a normal stellar weak field which is about one Gauss. Similarly, the immense rotation of a neutron star comes from the conservation of the angular momentum (mass times the angular velocity) of the original star. Both these features arise during the gravitational collapse of the evolving star past its equilibrium configuration. The external magnetic field captures electrons and ions, that produces a plasma magnetosphere (just as the magnetosphere of the Earth formed from the cosmic ray particles trapped in the Earth's dipolar magnetic field). Such a rotating plasmasphere radiates (accelerating charged particles radiate) a beam of electromagnetic waves which when intersects the lines of sight of observers on earth produces a pulse as received by earth recorders once in every revolution of the star. The discovery of these celestial clocks boosted research in observational radio astronomy immensely and also it encouraged physicists to work on varied approaches to understand the stability of such stellar configurations and their equations of state (relation between pressure, density, temperature and any other physical characteristic that govern the star) which is a fertile field for research. After the confirmation of the existence of radio pulsars, astronomers have discovered similar pulsars in other electromagnetic frequencies too like X-rays, and optical radiation. Also, some pulsars have been found in binary systems and these provide the most impressive arena for testing theories of gravitation and in particular for the theory of general relativity.

As a graduate student in radio astronomy, Ms. Bell along with a few colleagues had built a special antenna, to survey for new celestial radio sources in particular for the ones that are scintillating. As there were no advanced computers (1967), Ms. Bell had to analyze the paper recordings of the radio output (running for almost kilometers) for several days and in the process discovered one signal from the same source that repeated itself every 1.33 s. As no one, including her supervisor A. Hewish, believed the source to be celestial, she examined the recordings made earlier of the same part of the sky, a task that required intense observational skill and patient analysis, for days on end. Finally, she confirmed the celestial origin of the signal after finding three more similar signalling sources from a different part of the sky. This was in 1967 and in February 1968 she published the discovery (Hewish et al., 1968) of the first Pulsating Radio Source (PSR B 1919). The discovery of Pulsars opened up an entirely new aspect of radio astronomy that eventually lead to a new enquiry to the physics of compact objects as they are considered to be rotating neutron stars with associated plasma-sphere. Further the discovery of gravitational wave signals from colliding neutron stars which came in 2017, opened up a new branch of multi-frequency astronomy and is expected to be a fertile ground for new research avenues.

Experimental studies were being carried out by individuals all through 1700 and 1800 to understand the physical and chemical properties of bulk matter as well as different forms of energy like heat, light, and sound their nature, propagation, and associated features. With the advent of spectroscopic studies of elements and compounds in 1800, distinctions were made between the atomic and molecular spectra. The exact nature of emission or absorption of radiation of different frequencies was an interesting but unanswered question. The beginnings of the twentieth century saw a revolution with Max Planck's idea of the energy quanta and Einstein's identification of the 'photon' as the light quanta brought in an era of a very dedicated search for the theoretical understanding of spectral lines and atomic structure. While experiments of J. J. Thompson (1897) and of R. A. Milikan (1906) revealed the existence of electrons within the atom, (and the charge associated with them) it was Rutherford's experiment with scattering of alpha particles that revealed the existence of the central nucleus with a positive charge (1911). Following these discoveries came a surge in theoretical developments of quantum mechanics proposed and worked out by Bohr and Sommerfeld. The theory of atomic structure developed by these stalwarts along with colleagues and students explained both the emission and absorption of radiation energy by atoms while exhibiting their spectra. Apart from rendering the foundations of modern physics, this gave impetus to astronomers to analyze the spectra obtained of the radiations emitted by cosmic objects like stars and nebulae. Between 1912 and 1940 the progress in understanding the physics and chemistry of subatomic systems grew almost exponentially due to the continued efforts, both experimental and theoretical, and this had a direct influence on developments in astrophysics, as well as in life sciences including chemistry.

Edwin Hubble's discovery of the Cepheid variable (known as V1) in the Andromeda constellation (which was earlier thought of as a nebula) through identification of individual stars (1924) helped in getting a measure of the distance of the nebula and confirmed it as a separate galaxy outside the Milky Way. Similar analysis of many other nebulae helped in confirming their identification as distant galaxies and associating the distance estimates with the redshifts in the spectra of these galaxies (obtained through the efforts of V. Slipher (1929) established the famous Hubble law (velocity of galaxies moving away from us being proportional to their distances) which even today is the standard measure in describing the expanding universe scenario. It is interesting to note that repeated continuous observations of the Cepheids using the Hubble space telescope have confirmed these conclusions beyond doubt. With the addition of efforts to measure precise distances to far-flung galaxies by the Hubble space telescope, an essential ingredient needed to determine the age, size, and fate of the Universe was realized. It is important to note the contributions of Henrietta Leavitt (1910) who initiated the brightness measurement of the Cepheids and used them as reliable markers for astronomical distances. Then came the suggestion of G Lemaitre (the Belgian priest 1927) regarding the expanding universe from a primeval atom that initiated the idea for the Big bang theory of the Universe.

With the understanding of the theoretical developments, special relativity, and quantum mechanics, and their application, the atomic structure was understood clearly, and this lead to probing the nucleus further to reveal the more fundamental particles. Apart from electrons and protons, the discovery of neutrons and the neutrino initiated serious attempts to look for elementary particles that form the structure of matter. With the advent of high energy accelerators, the particle spectrum got established and paved the way for the development of the standard model in particle physics. Amidst these developments, the understanding of nuclear fusion led the way for describing the energy source of the Sun and the stars which brought a revolution in astrophysics regarding stellar evolution and the resulting discovery of neutron stars (pulsars) and black-holes.

One of the early discoveries in the field of condensed matter studies was the 'Hall effect' (1879), which is "the production of a voltage difference across an electrical conductor perpendicular to an electric current in the conductor and a magnetic field perpendicular to the current in the conductor". However one had to wait for a long time before explaining this feature, till Landau developed the theory in 1930 using quantum mechanics. An important property of magnetism (that was already known through the works of Faraday, Maxwell, and a few others) was found to be linked to temperature by Pierre Curie (1895) who discovered what is called the 'Curie point' in ferromagnetic materials when there was a phase transition due to material's intrinsic magnetic moments changing direction. Further work on this aspect saw the microscopic description of magnetism by W. Lenz and, E. Ising. These various studies seem to have laid grounds for the development of magnetic materials that could be used as storage devices.[1]

A very important development in this line of experimental research in condensed matter physics, dealt with the concept of 'superconductivity', which if realized at room temperature would be a big boost to the power industry as transmission lines for electricity without any loss could be achieved. Unlike a normal metal conductor whose resistance decreases only with lowering of temperature, a superconductor has the property of reducing the resistance to almost zero below a critical temperature. This was discovered by Kamerlingh Onnes (1911, who also was the first to liquefy Helium in 1908), using quantum mechanics and is characterized by the Meissner effect which ejects all the magnetic field lines from the interior of a superconductor. The theory of superconductivity was first explained by Bardeen, Schiffer, and Cooper (1957) using the idea of condensation of Cooper pairs, which incidentally also is considered for explaining the pairing of nucleons in an atomic nucleus. Cooper pairs are the result of a bound state of two electrons having opposite spins (like spin up and spin down) mediated by phonons (the fundamental quantum of vibrational energy coming from oscillating atoms) in the lattice of a crystal. High-temperature superconductivity was a largely discussed topic in the following decades and in 1986 Mueller and Bednorz discovered the first example of super-conductivity at 50 °K when it was also realized that this feature of high-temperature superconductivity resulted from strongly correlated material where the electron-electron pairs have an important role.

Studying the properties of matter using different probes like scattering by optical or X-ray photons and neutrons one could get information about the surface as well as

---

[1] https://en.wikipedia.org/wiki/Condensed_matter_physics

the interiors down to atomic scales. Such a study was identified as solid-state chemistry or condensed matter chemistry dealing with the understanding of intrinsic properties of materials such as chemical composition and structure. These studies could also result in finding new materials with tunable properties having applications in catalysis, energy storage, and different types of sensors with specific materials like metals, semiconductors, magnetic materials, and zeolites. The most interesting of the new material research is the possibility of having 'Bose-Einstein condensates' which is sometimes referred to as the fifth state of matter (after solid, liquid, gas, and plasma). As is known, among the elementary particles, the Fermions (having half-integer spins which follow Fermi-Dirac statistics) have to satisfy Pauli's exclusion principle and thus are forbidden from two of them occupying the same quantum state. On the contrary, the Bosons (particles with integer spin and satisfy Bose-Einstein statistics) can group and form a clump of atoms, entering the same energy state at very low temperatures. Such configurations were formed in a laboratory for the first time in 1990 using techniques of laser cooling and evaporative cooling. In a gas, one starts with a disordered state with kinetic energy greater than potential energy and when it is cooled down to extremely low temperature it doesn't form a lattice-like solid but all the atoms fall into the same quantum state and get indistinguishable behaving almost like photons. [There is an interesting story about Dirac worth noting. Apparently during his India visit in early 70s along with his wife, Professor Bose invited him to lecture in Kolkata. After the talk to return to their hotel, Professor and Mrs. Dirac sat in the back seat of a big limousine whereas Professor Bose and all his associates crowded in the front seat. It seems Mrs. Dirac being surprised at this odd distribution (only two of them in the back seat but many crowding the front seat) asked her husband why it should be so and it seems Dirac smiled and quipped "Statistics, my dear sheer statistics".] Though Einstein had predicted the possibility of such a formation based on Bose's work concerning photons, its realization had to wait for the development of technology of cooling, till in the 90s when C. C. Tannoudji, S. Chu and, W. D. Philips (Nobel Prize 1997) perfected the technique of laser cooling which can localize the atoms by slowing them down after confining them magnetically without a container. Using these techniques Cornell and Wieman (1995, Nobel prize 2001) succeeded in localizing about 2000 atoms of rubidium gas in a time-averaged orbiting potential trap for about 15–20 s. The techniques they adopted were further improved and a very low temperature of about 3×10^{-9} K has been achieved.

It is envisaged that BECs will have a very important role in the advancement of 'quantum computing' which has been a very ambitious project in the experimental science of the present century.[2]

Medical science, particularly human physiology has been enriched by numerous experimental discoveries almost from the eighteenth century, with experiments by Francois Magendie on dogs, aimed to understand the function of motor nerves and

---

[2] Studies in BEC have been very interesting and application oriented. See. https://www.nist.gov/news-events/news/2001/10/bose-einstein-condensate-new-form-matter

spinal nerves. Claude Bernard (1813–1878)—the father of experimental physiology explored the process of digestion through fistulas and the role of the pancreas and glycogenic function of the liver. These studies were pioneering and laid the foundations for various branches of medical science. German physiologists J. Hitzig and G. Fritish were the first in 1870 to show that different parts of the brain control different functions while the Russian physiologist Ivan Pavolv (1849–1936) is acclaimed for his experiments with automatic reflexes and animal behavior. His Nobel prize (1904) winning work investigated the physiology of digestion and the autonomic nervous system. He is also known for his work on conditioned reflexes (known also as Pavlov dog) where he established that the ringing of a bell could stimulate the dog to salivate (Spangenburg & Moser, 1994). These scientists are credited with the development of the branch of 'Neuroscience', which was initiated through the works of P. Broca who had suggested (1861) that any damage to the left frontal lobe of the brain could affect the capacity for the use of language, and the British neurologist J H Jackson's investigations revealed that different areas of the cerebral cortex controlled the movement of different parts of the human body. On the microscopic front, understanding the molecular biology of the human body is necessary for the development of diagnostic methods and associated tools and possibly finding drugs. This requires deeper knowledge of the chemical bonds and structures of both natural and synthetic compounds. As mentioned in National Research Council (2003) in the context of the Biology-Medicine interface, among important biomolecules of significance for possible medical applications are the polysaccharides whose chemistry has been a major topic of research. Though these are the most abundant biopolymers and can perform several structural, and regulatory functions their biological activity depends on their structure. As these are far more difficult to synthesize than proteins and nucleic acids, at the beginning of the present century (2001) automated synthetic methods were deployed. Continued research activity on these compounds seems to have shown that chemical modifications can significantly increase the diversity of its biological applications (https://www.bioglyco.com/). The study also has revealed that "a major lesson learned from the first draft of the human genome project is that there are lesser number of genes than originally thought which was a surprise. If the number has not increased with complexity how are the complex functions characteristics of higher mammals programmed? The answer seems to be in the post translational modifications of proteins achieved with added features. It was surmised that a single protein derived from a single gene can be transformed into numerous distinct molecular species and thereby increase the information content of a very concise genome which are more extensive in higher organisms".[3]

Starting from environmental studies, spectroscopy in different wavelengths (IR to X-rays) of irradiated material samples either through their identification in liquids and solids or through atomic absorption spectrometry has been in practice. IR and

---

[3] For an extensive literature related to the topic 'biology and medicine frontier' see http://nap.edu/10 633

UV spectrophotometry of in situ sites in the atmosphere through remote sensing have given a lot of understanding on the environmental changes which have occurred both due to natural and human-influenced causes. X-ray fluorescence has been in practice for a long time helping to determine the composition of different solid samples both in agriculture (for soils) and industry. In biomedical research optical properties of tissue responses have been analyzed using the techniques of photon time of flight which has been useful in medical diagnostics also as they are non-invasive. Near IR spectroscopic methods in pharmaceutical applications have been very useful as they are non-destructive and require no sample preparation (Tomislav, 2019).

More recently the technique of LIDAR (Light Detection and Ranging) is used for remote sensing as well as for the measurement of Earth features. Its basic idea is to use a pulsed laser to measure different ranges to earth which along with other data (obtained from airplanes and helicopters) can give the three-dimensional information and surface features of the Earth, as it uses the special scanners and the coordinated GPS data. This technique has been very useful in hazard assessments like lava flows, floods, landslides, and tsunamis apart from its role in geological mapping and river surveys. More importantly, the high level of resolution made available by lasers help in analyzing the particulate matter in the atmosphere which can help in pollution control and other environmental studies (AGI21).

Among the famous experimenters in science in the Indian context, the names of J. C. Bose (for Physics and Botany), P. C. Ray for his work on chemical compounds of mercury, and C. V. Raman (for Raman effect) have been of great significance. Bose and Ray were contemporaries (both were students at the same college) and pursued their fields of interest with great alacrity and vigor, though lacking in funding and equipment. Bose had learned of Maxwell's theory of electromagnetism (1865), the equations of which had wave-like solutions, and thus discovered a method to produce those waves (microwaves of wavelength 5 mm) as well as to detect them at a distance using semiconductor material.

Unfortunately as he was not keen on patenting his discovery, two years later Marconi, the Italian engineer reinvented the same and received the credit for the discovery of wireless communication. This clearly shows how important it is to make public (announce formally) any discovery one makes in science or technology. J. C. Bose is a fine example to emphasize the integrated approach to science, as he worked on topics both in physics and botany, particularly while analyzing the growth and reaction of plants to external stimuli. This he studied by constructing automatic recorders capable of registering extremely slight movements and quivers in plants.

Prafulla Chandra Ray, known as the father of chemistry research in India, and is acclaimed for his work on mineral salts (sulfates and nitrates), discovered Mercurous nitrite (in 1896), a stable compound, which later crystallographic techniques proved to be novel and correct. Among other important studies of Ray, the synthesis of ammonium nitrite in pure form using the double displacement technique between chlorides and silver nitrite is recognized widely. His research methodology and guidance attracted several students who worked with him and later established several schools in India (BSS 2017; En Wikipedia, 2021).

C.V. Raman's discovery of the 'Raman effect', is one of the most important studies in the analysis of molecular spectroscopy. The discovery [(1928), Nobel prize (1930)] shows that when a beam of colored light passes through a liquid, a fraction of it gets scattered by the molecules of the liquid to a different color. The pattern of the Raman lines is characteristic of the particular molecular species and the intensity of the scattered light is proportional to the number of molecules in the path of the ray. This result named after Raman was found to have innumerable applications, and nowadays with advanced technology coming from computers and lasers, the applicability has increased immensely.

Over the last fifty years, experimental research has taken lead over the theoretical efforts. Most advances have been in life sciences with applications in the areas of genetics, genetic engineering, environmental sciences, etc. In the physical sciences high energy physics, astronomy, and cosmology (due to the advancements in space technology) have seen substantive progress and continue to do so as they are well supported by the advancements in electronics, nanoscience, and computer science. The most important discoveries in astrophysics and cosmology happened with the discoveries of very high energy sources like Quasars, AGNs, and Pulsars. Detection of the Microwave background radiation and finally the detection of Gravitational waves reaching the Earth were the most important breakthroughs. The experiments and observations in this regard are mostly efforts of large groups of scientists and greater cooperation between industries and institutions along with governmental support in funding these efforts.

Experimental science got a big boost with the advancement of precision technology as evidenced both in physical and biological sciences. The precisions achieved were not only in instrumentation but also in speed and accuracy in computation. In high energy physics, the standard model of particle physics has been in vogue for almost 50 years, which gives a model for a unified treatment of the electromagnetic interaction and the weak interaction responsible for radioactivity (electroweak theory), combined with quantum chromodynamics describing strong interactions. As a quantum field theory, the model describes both the matter particles like electrons also known as fermions and the interaction carriers like photons also called bosons, as the underlying quantum fields. The theory requires certain symmetry to be broken as described by Murray Gellman (1964), which introduced the requirement of the force-carrying particles to be massless. As far as electromagnetism was concerned this was satisfied (photon is massless) but the carriers of weak force the W and Z bosons were not. To overcome this problem, Brout, Englert, and independently Higgs had proposed a new mechanism for breaking the electroweak symmetry by introducing a new quantum field the Higgs field, whose quantum manifestation was later called the Higgs boson. This provided a solution as the mass for W and Z bosons was attributed to their interaction with the Higgs field. Though the problem appeared to be solved theoretically unlike photon the carrier of the electromagnetic force the Higgs particle had not been found in any of the high energy experiments. Looking at the history of elementary particle physics, it was known that almost all particles discovered during the 1950s and 60s were basically predicted on theoretical reasoning, and also one had to build new accelerators with increasing energy regimes for their discoveries. As the new and higher energy accelerators produced enormous data of high energy collisions, analyzing the data took a lot of time and effort which increased the particle physics community in the 70s manyfold. As the governmental support reduced for building accelerators, one saw in the 80s a surge of particle physicists moving towards astrophysics and cosmology in search of higher energy regimes in the early universe scenario. This was with the hope of finding new particle signatures that could ensure the existence of the Higgs boson apart from looking for more fertile grounds for research. However, very concentrated and continued efforts by the two groups, the Compact Muon Solenoid (CMS) and ATLAS (a general-purpose detector at the Large Hadron Collider at CERN), mainly concentrating on the precision and complex data acquisition and computing systems finally succeeded in identifying the elusive 'Higgs particle' (some people called it God particle) in 2012 around the mass range of 125 GeV. (It is said that more than 3000 scientists from 174 institutes belonging to 38 countries worked on the ATLAS system.). As mentioned by one of the co-conveners of the program discovering the particle at a convenient mass was an unexpected kindness from nature. However one had to treat the uncertainties similarly across all the individual analyses and interpret the results carefully through sophisticated techniques after going through the statistical analysis. Combining data from the transformation of the Higgs boson to pairs of Z bosons and pairs of photons allowed ATLAS and CMS to the discovery (Rao, 2020). The efforts and the dedication of scientists in discovering Higgs can be matched only to the efforts put in for detecting gravitational waves by LIGO and VIRGO collaboration in 2015.

Unlike in physics, several new discoveries in life science research may not be the consequence of well-established theories. Most of the efforts which opened up new lines of investigations have come out of individual efforts whose work got noticed and extended by others to find new applications. Gene therapy which was considered as fiction however got expression in 1972 when Friedmann and Roblin (1972), advocated the idea of developing techniques for gene therapy with caution against using human patients. However, by 1990 clinical trials got approved on a patient with a severe immune system deficiency, and soon after cancer gene therapy tests were conducted. Today gene therapy is an important strategy for treating disorders caused by either missing or a faulty gene-corrected by adding, inhibiting, editing, or by functional replacement of a gene. The utility factor in this method is mainly that it is designed to be a one-time treatment that targets the root cause of a disease. In this context, an earlier breakthrough by F Sanger and colleagues in 1977 regarding the rapid sequencing technique of DNA came in handy for the human genome project that got launched in 1990 and was completed in 2003. This was a huge project that involved massive sequencing and computing power in 20 institutes across six countries with thousands of individuals and considered as the largest collaborative project in life science research.

Developments in molecular biology found a big boost with the discovery of 'RNA interference' by A. Fire and C Mello in 1998, the work that revolutionized the understanding of the processes and regulation in molecular biology (Shaver, 2018).

By and large scientific research has progressed over the last century and a half, from individual efforts to collaborations necessitated both by needed resources and personnel. This has had some effect on the social behavior of scientists involved in large projects as they cannot discuss in public any of the results they may come across, and obligatorily have to share the findings with all colleagues in the project. Sometimes this could be unfortunate but unavoidable. This can have some effect on some practitioners who will have to take proper care not to get disappointed or disenchanted. However, this problem is more in experimental areas and far less in theoretical studies.

Doing experiments to either support or reject a paradigm is similar to what one does in several normal situations in daily life-try and find out. As Kuhn (1962) points out there seem to be three normal foci for factual scientific investigation which are neither always nor permanently be distinct. The first of the foci deals with facts that can show that the paradigm reveals the nature of things and by using it one can consider it important to do the supporting experiments with more precision and for a larger variety of situations. Attempts to increase the accuracy and scope with which facts regarding such paradigms hold are the most frequent both in experimental and theoretical investigations. The second focus deals with experiments to check the predictions of paradigms like the famous expedition of Eddington to observe a total solar eclipse in 1919, to check Einstein's prediction of the bending of light ray passing the limb of the sun. This is a very important aspect of scientific methodology because should the prediction comes out to be not right then one may have to look for a new theory either changing the assumed hypothesis or through checking the logic of the prediction which might have missed some information. The third focus consists of empirical work undertaken to articulate the paradigm/theory resolving some of its residual ambiguities and permitting the solution of problems to which it had previously just drawn the attention to. He emphasizes particularly the fact that often a paradigm developed for one set of phenomena could be ambiguous in its application to other closely related ones; when new experiments may become necessary to choose among the alternative ways of applying the paradigm to the new area of interest. An important feature that needs careful consideration is a situation that may arise during a routine set of experiments when something completely unexpected and unpredictable event occurs to draw the attention of the investigator. A famous example of such a possibility was in the discovery of X-rays by Röntgen. He had been studying the properties of cathode rays for some time when one day he accidentally noticed that a barium platinocyanide screen at some distance from his shielded apparatus glowing while the discharge was in progress. He suspected that the glow came from the cathode rays being stopped by some agency and made the screen glow. Immediately he changed his earlier goals and worked continuously on the new phenomenon (radiation coming off the metal screen from the impact of cathode rays) and realized that the new radiation cast shadows, and was not affected by the magnetic field. After several hours of repetitive experiments, he was fully convinced that the new radiation though is similar to light appeared to be of higher energy that it could pass through several routine blockades which were opaque to light. He called the new radiation X-rays which later came to be known also as Röntgen rays.

As Kuhn elaborates, it goes to show how experimentation is a basic approach for scientific investigations that involve conscious manipulation of certain aspects of a real system and the observation of the effects of that manipulation. Many times it helps to determine the relationship between dependent and independent variables of the experiment if the experiment is performed with slightly different procedures with minute modifications. It is known that exploratory experiments are carried out many times in medical biology to determine the dosage strengths of drugs for different diseases for different patients. Under such investigations, it is very necessary to keep in mind possible undesirable effects that may result in the context of such experiments, particularly in the trial runs of newly discovered drugs. Quite often several animal species (particularly white rats and rhesus monkeys) are sacrificed in several such experiments. This ethical issue must always stand in guidance to investigators who in their enthusiasm may not think about the consequence.

As is very well known unfortunately the invention of the atomic bomb that was used in the second world war was indirectly an effect coming out of Einstein's most famous mass-energy relation $E = mc^2$, and the discovery of fission chain reactions by Szilard and Fermi. Though on the one hand, the understanding of the equivalence of mass and energy has been one of the most significant results of modern science its usage in producing the weapon of mass destruction has remained a black mark on research in science. However such unfortunate consequences coming mainly from the governing politics and its supporters in the society raises important ethical issues about applied research. On the positive note, one should remember more about the useful consequences of experimental research which certainly surpass the ill effects and help in understanding the paradigms under investigation leading to further progress...

References

AGI21. https://www.americangeosciences.org/critical-issues/faq/what-lidar-and-what-it-used
Breakthrough Science Society. (2017). *Brief history of Science*.
En Wikipedia. (2021).
endesa17. https://www.endesa.com/en/blogs/endesa-s-blog/others/who-actually-invented-the-light-bulb also https://www.bulbs.com/learning/history.aspx
Friedmann, & Roblin. (1972). *Science, 175(4025)*, 949–955. https://doi.org/10.1126/science.175.4025.949
Gold, T. (1968). *Nature, 218*, 731.
Hewish, A., Bell, S. J., Pilkington, J. D. H., Scott, P. F., & Collins, R. A. (1968). Observation of a rapidly pulsating radio source. *Nature, 217*(5130).
https://en.wikipedia.org/wiki/Jagadish_Chandra_Bose
https://en.wikipedia.org/wiki/Prafulla_Chandra_Ray
Klepner, D. (2007). *Physics Today 60*, 8. https://doi.org/10.1063/1.2774115

Kuhn, T. (1962). *The structure of scientific revolutions*. University of Chicago Press.

Michelson, A. A. (1879). *American Journal of Science, s3–18*(107), 390–393. https://doi.org/10.2475/ajs.s3-18.107.390

National Research Council. (2003). *Beyond the molecular frontier: Challenges for chemistry and chemical engineering*. The National Academies Press. https://doi.org/10.17226/10633. http://nap.edu/10633

Prasanna, A. R. (2008). *Space and time to space-time*. Universities Press.

Rao, A. (2020). *The Higgs boson: Revealing nature's secrets*. https://home.cern/news/series/lhc-physics-ten/higgs-boson-revealing-natures-secrets

Spangenburg R., & Moser, D. K. (1994). *The history of science* (in the Nineteenth Century) (91–95).

Shaver, P. (2018). *The rise of science*. Springer.

Tomislav, M. (2019). *Spectroscopy applications*. News-Medical. Retrieved November 07, 2021, from https://www.newsmedical.net/life-sciences/Spectroscopy-Applications.aspx

Chapter 8
Observation and Inference

Introduction

The beginnings of science as is known today started with thinkers who started wondering about how and why of the events around and the quest to understand our origins. It is the developed capacity to observe natural events, like the motion of the celestial bodies in the sky, the growth and the varieties of living beings, movment of animals and birds, and attempts to understand the causes in each such case that motivated the pursuit of science. Development of the skill for questioning, reasoning, hypothesizing, and verifying the cause-effect relation became a part of the developed human brain which automatically led to applying the understood phenomena to inventions and discoveries that formed the basis for technology. Famous examples in this direction can be seen in the lives of T. Brahe, J. Kepler, G. Mendel, K. Jansky, A. Fleming, J. C. Bose and many other similar personalities. Meticulousness in setting up experiments and recording observations systematically are the hallmarks of a good scientist. As a process of creative thinking, practicing science depends upon the four stages; preparation (to hypothesize), incubation (of the idea for deeper thinking), illumination (finding results based on the ideas), and verification (reproducibility of results). These elements are cardinal for a successful pursuit of any discipline of intellectual endeavor and particularly of science.

Man's curiosity started with his desire to understand the nature and events around him. Though natural phenomena occurred as they should in their periodicity, it was the perseverance and the capacity of a few to observe them that led them to discern the patterns that exist in Nature. Further observations led them to understand the modes of occurrence of different natural phenomena, which at times saved them from the impending dangers like having readymade disaster management aids before unavoidable natural events happened.

Some of the thinkers perceived the necessity to make a continued observation, particularly of the night sky, to understand the periodicities in the motions of celestial bodies, and were bold enough to suggest a set of natural laws, which

© The Author(s), under exclusive license to Springer Nature Switzerland AG 2022
A. R. Prasanna, *How to Learn and Practice Science*,
https://doi.org/10.1007/978-3-031-14514-8_8

were mathematically rigorous and consistent with observations. Though the early attempts of Greeks in describing the cosmos and its structure, were later found to be incorrect, approaches developed by them provided a useful methodology for doing and communicating science.

Observations play a crucial part in understanding both the macro and the micro worlds. Humans cannot create or control natural phenomena, but surely can try and understand them through repeated observations. With sufficient data, one then models the phenomena through mathematics and constructs a theory based on specific postulates. The earliest attempts of such a methodology can be traced to Tycho Brahe and Kepler, the pioneers of the discovery of celestial mechanics—the laws of planetary motion. Their contributions were based on painstaking observations and meticulous inferences, that eventually supported Newton's discovery of the law of gravity. Whereas the contribution of these pioneers gave a foundation for understanding the planetary orbits, it is the discovery of 'stellar parallax' that established a technique for finding their distances and of nearby stars. The technique is simply to measure the star's apparent movement against the background stars at larger distances, as the Earth moves around the sun in its orbit. The method worked efficiently for the nearby objects with a parallax of more than 1 arcsec. It appears that after almost a hundred years of these measurements it was only in 1838 that Wilhelm Bessel reported a convincing measurement of 0.28 arcsecs for the star 61 Cygni which implied a distance of almost ten light-years. Measurement of parallaxes less than 0.01 arcsec had to wait for telescopes with the Earth-based ones going down to about 0.01 arc secs while the space-based telescopes of the modern era reach as low as 0.001 arcsecs which is still limited to stars in our galaxy. For far-away stars, the distance measurement is done by using the brightness variation of the Cepheids and for faraway galaxies, the brightness variation of Supernovae is used. Among the discoveries made through keen observation of interrelationships of stellar features, the H-R (Hertzprung-Russel) diagram (1913) is very important. This depicts the connection between the luminosity (intrinsic brightness) and the color of stars. Hertzprung in 1911 and Russel in 1913 had independently discussed the relationship between the spectral class of stars indicated by their temperatures expressed as colors (black body relation) and their luminosities. A typical H-R diagram is shown in the adjoining figure (Fig. 8.1).

The diagram clearly illustrates the distribution of stars belonging to different regimes of evolution. The central portion running from top left to bottom right is called the main sequence which show mostly middle aged stars. It seems to show that the hotter stars are more luminous than the cooler ones which perhaps is expected. As marked on the graph, one sees that there are a number of stars above the main sequence which are high in luminosity but of lower temperature. One may ask how could this be? These are indeed referred to as giants and super giants as their sizes are so enormous that their total integrated energy output is large and thus are more luminous. Similarly one finds the opposite a group which are of higher temperature but of lower luminosity indicating that their total surface area is smaller and consequently of lower dimensions and thus are called the dwarfs. This diagram

Fig. 8.1 H-R diagram credit: Encyclopedia Brittanica inc. Britannica, The Editors of Encyclopaedia. "Hertzsprung-Russell diagram". *Encyclopedia Britannica*, 8 Jun. 2021, https://www.britannica.com/science/Hertzsprung-Russell-diagram. Accessed 1 June 20 22

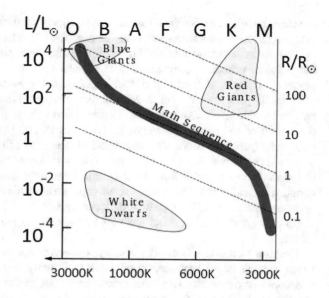

has been very useful in astrophysics particularly for the study and analysis of stellar evolution.

The first instrument for seeing objects at far away distances was devised by a Dutch spectacle maker H. Lipperhey in 1608, and on learning about it Galileo made a similar instrument using lens that he thought would help in seeing early the enemy ships from far away and thus be useful in the defence of his country. However, later when he turned the instrument towards the night sky had the inspiration to discover the four primary moons of Jupiter, observe the rings of Saturn, and, apart from these also noticed Sunspots and the lunar surface endowed with ridges. Kepler was the next to build a telescope with improved optics and also explain its working, using the principles of refraction of light. Next came the telescopes of C Huygens and I. Newton. While Huygens ground his lens and constructed a 12-foot telescope that helped him discover Titan the moon of Saturn, Newton deviated from the rest to build a reflecting telescope using mirrors instead of the lens. Using mirrors he could avoid any distortion due to chromatic aberration (failure of different wavelengths of light coming to a common focus) and the first reflecting telescope was built in 1668. Further developments helped observation of the cosmos and laid the path to astronomical progress as well as enlarging the view of the Universe. The 36-inch refractor at the Lick observatory on Mt.' Hamilton, California seems to be the largest optical telescope in operation. However, reflectors with improved mirror designs and optical paths of the incident beam helped observations of deep space objects. Difficulties in making larger mirrors and desire to increase the light-gathering power brought in the idea of multi-mirror telescopes, the first such being the two 10-m combinations in Mauna Kia (Hawaii) belonging to the Keck observatory which came into operation in 1996. Each of these consists of 36 contiguous adjustable mirror segments operated by computer control. Along with this several improvements were brought in for the auxiliary instruments like the photomultiplier, spectrophotometer, and CCD (charge-coupled devices) which increased the 'seeing' facility. The peak of achievement in optical telescopes came with the 'Hubble space telescope' put in orbit in 1990, which has a 2.4-m primary mirror along with several instruments like a wide-field and planetary camera, a faint-object spectrograph, a high-resolution spectrograph, a high-speed photometer, and a faint-object camera. While the observations of Edwin Hubble made from the Mt.' Wilson observatory in the 1930s put forward the notion of the expanding universe, HST has helped in determining the rate of expansion (probably more accurately) and set the age of the Universe to be at 13.8 billion years. It has also enabled one to take very sharp pictures of distant cosmic objects as it avoids the distortions that come from the atmosphere, and has provided conclusive evidence for the existence of supermassive black holes at the centers of galaxies. From the point of view of the general public, HST has provided brilliantly sharp photographs of stars and galaxies that reveal the vastness and beauty of the Universe.

Coming to the late twentieth century, observational astronomy got a very big boost because of the advances in optics as well as the introduction of space-based telescopes covering almost all the wavelengths (IR to Gamma rays) adding to the already established Radio astronomy (1940). IR astronomy owes its origin to the experiments by W Herschel who in 1800 tried to measure the temperature of different colors of the solar spectrum and found that the temperature was the highest at the IR wavelength. As the water vapor in the atmosphere absorbs the IR, it was felt that trying to observe from mountain tops might be more helpful. Accordingly, E. F. Nichols trying from Tenerife though was not very successful, his report of the flux ratio in IR of the two stars he observed agrees with later observations and thus he is credited with the first detection of IR from stars. Improvements in IR detection technology encouraged astronomers to look for IR emitters in the 1960s and the first all-sky survey of IR objects happened only in 1983 with the launching of IRAS (Infra Red Astronomical Satellite) and later in 2003 with the Spitzer space telescope that helped detect emission from Double Helix nebula and extrasolar planets. Added to the Radio and Optical cosmic sources, IR astronomy gained importance, and the inputs from all these made astrophysics and cosmology a very fertile ground for researchers from physics and astronomy communities. It is indeed a marvel that the new generation has planned and executed a wonderful instrument in designing and keeping in place the James Webb space telescope at the second Lagrange point between Earth and Sun. This multi mirror telescope which has been deployed in January 2022, will start giving its observational results by September 2022. Using the advantage of the extreme cold conditions where it is located, Webb will concentrate on observation of objects at far-IR wavelengths that could indicate the very early stages of a forming star and also try to get the spectra of exoplanet atmospheres which could contain water vapour, carbon dioxide and methane which perhaps are indicators of life. As Webb will be able to see far-IR wavelengths it could penetrate the space-time to a far greater depth than what has been achieved till now and that would boost our understanding of the very early universe.

With advances in space technology using satellites and space probes, astronomical research went a step further to include the other bands of the electromagnetic spectrum as one could now see UV, X-ray, and Gamma-ray emissions from the deep space thus extending the view of the Universe. Satellite-based observation of UV emission identified mostly the early or late stages of stellar evolution (the hot phase). An international collaboration (NASA, UKSERC, and ESA) Explorer 57, proposed in 1964 but launched in 1978 lasting for 18 years provided several interesting findings like (1) the rate at which comet Halley lost its dust and gas, (2) a large number of hotter stars (with temperature > 10,000 K) which emit most of the radiation in UV, (3) White dwarf companions of many main sequence stars, (4) evolutionary history of binary stars and (5) the most interesting Supernova 1987A. IUE was also very helpful in understanding the AGNs particularly the Seyfert galaxy and the absorption lines in quasar spectra that revealed the amounts of Hydrogen clouds between Earth and the Quasar. Comparison of the spectra of the nearby Quasars with those of distant ones (studied in the optical range due to redshift from expansion) has helped in understanding the density of hydrogen clouds in different regions of the extragalactic space possibly revealing the fact that the clouds have turned into galaxies over time as there seem to be fewer clouds in the nearby region as compared to far away distances. The UV spectral input has increased for astronomy with the HST with its additional aid Cosmic Origin Spectrograph installed in 2009 that can record an ultraviolet spectrum of objects looked at.

Even though X-rays from the sun were recorded by sounding rockets since 1958, the first source outside the solar system was the discovery of Scorpius X-1 whose X-ray output was found to be about 10,000 times greater than its visual emission. With the launching of the satellite UHURU in 1970 by NASA the first sky survey in X-rays was conducted as well as it discovered a whole set of objects ranging from binaries to supernova remnants and diffuse emission from galactic clusters. The main contribution from the X-ray sky was the identification of highly collapsed stars (neutron stars and black holes) which helped a large-scale effort from astrophysicists to understand the stellar evolution and death in various manifestations. As a tribute to the theoretical efforts for understanding the black hole physics, led by S. Chandrasekhar, a dedicated satellite with an X-ray telescope named Chandra was launched in 1999, which is considered as NASA's flagship as it carries a most sophisticated X-ray observatory. It has uninterrupted observation of the cosmos for over 55 h during its long orbital time of about 64 h as it spends about 85% of the time beyond the Earth's charged particle surroundings (von Allen belts). Starting with the picture of Cas-A (remnant of the supernova observed by Tycho Brahe) Chandra has revealed a ring around the Crab pulsar and several X-ray emitting stars in the Orion constellation. Early in its lookout, it studied the black hole candidate M82 and found enough evidence to model the object as a medium mass black hole having around 500 M_{sun}. One of the most interesting observations that revealed a new line in astronomical research was the detection of X-ray emission from the 2017 gravitational wave source GW170817, which seems to be coming from the collision of two neutron stars. This detection of X-rays, further augmented by signals in the other bands like the Radio, IR, Optical, and Gamma-ray opened up the doors for multi-frequency astronomy. This is perhaps the future as it will encourage collaboration among astronomers, physicists, and engineers from different parts of the world and different laboratories and help in putting together the resources and human power for the cause of science.

Another important development in the 1960s was the detection of Cosmic background radiation (CMB), discovered by Penzias and Wilson (1965) in the microwave region of the electromagnetic spectrum. Just around the same time when this discovery was announced theoretical cosmologists Peebles and co-workers had predicted the existence of the remnant radiation from the Big bang (predicted in 1948 by Gamow et al) at the observed temperature of about 2.7 °K. From an observational viewpoint, Penzias and Wilson had checked thoroughly all possible sources for the observed signals (including a remote chance of local disturbance from nesting birds) but had not found any probable source. With this in the background and learning about the prediction of Peebles, they associated their discovery with Peebles' theory confirming that the discovery was the Microwave background radiation produced during the era of recombination getting cooled over the expanding universe. This helped the growth of Cosmology as a scientific discipline which over the years has increased the knowledge about our universe.

Gravitational-wave astronomy which got established in 2015 with the discovery of the emitted waves by colliding blackholes, again comes under the second focus as suggested by Kuhn (previous chapter) and is a product of confirming a predicted result from an already successful theory-the general relativity. Einstein after publishing the theory of general relativity explaining gravity as a field had conjectured the possible existence of gravitational waves associated with the gravitational field similar to the electromagnetic waves associated with the electromagnetic field. He had worked out the possible solution to the linearized system of the field equations (1917) representing the gravitational waves. However, as there was no experimental facility sensitive enough to see these waves at that time it remained of theoretical interest only. Though not successful, one needs to mention the observational attempts made by J. Weber during the 1960s, for the technique he employed to detect gravitational waves. As nature has exhibited time and again, in 1974 Hulse and Taylor who discovered a pulsar with 59 ms periodicity also had noticed that there was a systematic variation in the arrival time of the pulses. Continued observations and analysis surprisingly showed the variation itself having a periodicity of 7.75 h. As was already known in astronomy such a variation is exhibited by binary systems with one object going around the other and this was confirmed as they realized that the pulsar is going around another neutron star. Thus came the discovery of the binary pulsar. Hulse and Taylor received the Nobel prize in 1993. Many important and interesting parts of this discovery came in 1980s when a systematic study of the recorded pulse arriving times year after year since its discovery was analyzed and the orbital parameters were determined. Both components of the binary were discovered to be neutron stars, of equal mass (about 1.4 solar mass). The orbit is inclined about 45 degrees with respect to the plane of the sky, and the relativistic precession of the periastron is about 4.2 degrees per year. Apart from these features, the most important news from the binary orbit was that the orbital decay year after year matched exactly with the loss of orbital energy due to the emission of gravitational waves as was required by Einstein's quadrupole formula (1917). Thus it was realized that the Hulse-Taylor binary was a perfect laboratory for checking the predictions of general relativity (Weisberg et al., 1981a, b). Continued observation of the various parameters as mentioned for another few years confirmed the findings and thus the prediction of the existence of gravitational waves was confirmed. However, attempts to discover them using an earth-bound laboratory were still due and the project LIGO got the required sanction. Now it was a team effort and a huge consortium of scientists (which included laboratory technicians, professors, research scholars, students, and a host of computer personnel) worked on increasing the sensitivity of the system (mainly to overcome all possible terrestrial noise levels). These efforts aided by new stable simulation computations by Pretorius in 2005 finally gave a successful result in 2015 when the first recording of the gravitational wave signal was achieved and analyzed as coming from the coalescence of a system of binary black holes of mass 29 and 34 M_{sun} resulting in a black hole of mass 60 M_{sun} at a distance of about 440 Mpcs from earth. The mass equivalent energy radiated as gravitational waves was estimated as about 3 ± 0.5 solar mass. Subsequently in the next few years, several other mergers have been detected and the most interesting of them was the coalescence of two neutron stars detected in Aug 2017 an event that gave birth to 'multi frequency astronomy' (as mentioned) from a single source as simultaneous observations of the event showed emissions in Gamma rays, X-rays, UV, Optical, IR, and Radio waves.

Developments in the other areas of science also have shown similar trends of experimentally finding some evidence to prove some already predicted effect or result. Darwin's theory of evolution of life which was based on his meticulous observations on the types and varieties of different species and their inter-relationships as well as Gregor Mendel's intensive work on 'peas' leading to the science of genetics, was confirmed and reconfirmed through further observations and inferences therefrom. In science, as there is always the possibility that the new observations may not conform to an existing theory the process of ideation and verification is slow, particularly when changes are brought in, and therefore progress in science is always slower, as it needs careful and multiple scrutinies at each stage of development.

As Shaver (2017) points out "the explosive development of biology as a fundamental aspect of life science research has resulted in a wide variety of new topics now being explored. A case in point is how the humans evolved to be different from other animals. The number of genes in our genome is far less than that of a 'wheat grain' and our brains are only about three times the size of those of chimpanzees. Obviously our superior brain power does not depend upon the size of either the genome or the brain. It could perhaps be due to the interactions and moderations of gene expression. Luckily the availability of complete sequences of the genomes of modern and ancient humans, chimpanzees and several other species of the animal world as well as of different organisms has evolved the study of 'differential evolution'. This has resulted in findings that while the DNA sequences are conserved in other animals they have changed in humans rapidly due to what are called Human

Accelerated Regions, which seem to enhance the modulators of gene expressions". These findings are of recent origin (2006) and HAR is a set of 49 segments of the human genome that are conserved throughout the vertebrate evolution but are strikingly different in humans. They were found by scanning through genomic databases of multiple species, which seems to have revealed that some of these highly mutated areas may contribute to human-specific traits while some others may represent the loss of functional mutations possibly due to the action of biased gene conversion rather than adaptive evolution (HAR21wiki).

The understanding of the basic principles in different branches of practiced science lead to the realization of the interdependence of the different sub-areas including their statistical and mathematical aspects. This has taken a major role in defining the research programs in the latter part of the twentieth century and the present century. Such studies revealed the type of questions that could be asked of several issues which were once thought as fully clarified. Two important problems that are glaring in the minds of physicists are.

1. quantizing gravity and looking for a fully unified theory of all fundamental interactions and
2. the basic difficulties associated with the physics of fusion (turbulence and nonlinear aspects) facing those involved (for almost 80 years) with research in controlled fusion as a source of usable energy.

There must be similar impediments in other areas too.

Implicit in the above is the fact that any theory that does not permit being questioned is by definition untenable. That said, it is also true that there is as yet no theory that explains all the facts within a given set of fundamental laws. This makes the quest for understanding the Universe, always a 'work in progress and this ensures that the creativity and innovativeness of the human mind remains an eternal active enterprise.

Observations lead to new theories and thence new knowledge. Observations relate every phenomenon to experience through sensory organs at times aided by instrumentation and technologies. Thus the earliest of human inventions to understand Nature were the telescope and the microscope. These aided the human eyes to see the largest, the farthest, and the smallest features of the Universe. Inference from observations is by itself a very creative process. It is argued that scientific creativity comprises a sequence of processes (stages) defined as, Preparation, Incubation, Illumination, and Verification (Patrick, 1955).

Preparation is the stage when after fixing the theme of investigation, the thinker, studies a little deeper about the problem to acquire more information. In this period the ideas shift rapidly and one's thoughts are not yet dominated by any coherent theme or formulation. At this stage, one needs to discuss with the peers and subject experts about his/her research problem to analyze more carefully the effectiveness of his/her methodology and 'modus operand'. Preparation involves both deliberate and non-deliberate mental activity. Sometimes, like in the case of a poet or an artist, ideas could seem to be pressing upon the mind without too much effort, whereas a scientist or inventor may spend hours in the strenuous mental effort to collect more information about his/her problem under investigation. This stage may often be accompanied by unpleasant feelings characterized by some doubt or perplexity. There have been cases when a sense of frustration may envelop the investigator, particularly when no solution comes in sight even after long periods of preparation. This stage may vary from a few hours to months or even longer and is followed by the stage of incubation.

The stage of incubation which many times may overlap the preparatory stage is characterized by the recurrence of the initial idea which may get repeated with modifications. It has been observed that many times the investigator might have considered changing his/her mode of thinking and moved on to some other activity like music, sports, or some physical exercise or just be relaxing. During this stage at times subconsciously, the incubating idea would get more clarified along with some new ideas that may develop and this process can be painfully slow or instantaneous if one is lucky. The incubation time is different for different individuals depending upon the age, experience, and the strength of the original impetus, as also the personal habits of the individual. At times these two stages of preparation and incubation could have overlapped even when the individual concerned is still in the process of gathering more information on the problem under investigation.

The third stage of creative thought is illumination when the incubating idea takes a definite form and gets expressed in varied forms in different pursuits. For example, a poet may write a few lines of a composed poem, whereas an artist may sketch the outlines of a painting. For a scientist, this is the situation when he/she realizes the glimpse of a solution. In all cases, such glimpses of a new idea accompany emotional happy outbursts like the famous 'Eureka' event of Archimedes. Normally it will be of short duration to be followed by a serious effort of consolidating the occurred idea and expanding it to a workable theory or model in science, or a new composition of a poem or a painting in art.

The final stage of the creative process particularly in scientific research is 'verification/confirmation' where the main idea is checked for its correctness in a scientific investigation and its aesthetic appeal for an artist or a poet. This stage which normally has to conform to the contemporary standards of the field of pursuit may require special techniques and could be more time-consuming than one had assumed. Keeping it slow and methodical would yield excellent results. A scientist may employ laboratory experimentation or statistical techniques of data analysis, whereas a musician plays on the composition on an appropriate instrument to see what notes/chords need to be changed, while an artist adds or eliminates a few lines or colors of the picture sketched.

The process of 'creativity' is thus a slow process with all the above-mentioned stages, with each one being different for different individuals and their pursuits. It is always possible that some endeavors get to be more productive (like in literature or arts) as compared to that of a scientist, but this is only because practicing science requires following conformity and applicability very rigorously. There could be discoveries leading to useful outcomes where the stages mentioned need not have been pursued but resulted just from keen observation.

An interesting example of such a discovery is that of the adhesive Velcro (Shaver, 2018). As is pointed out, a Swiss engineer George de Mestral walking his dog in the woods noticed (1941) that the weeds of cockle-burs (of genus Xanthium) had attached themselves to his clothes and the fur of the dog which made him realize that the adhesive nature was due to interlocking of tiny spikes of the burr to certain surfaces. Working with very tiny hooks fixed to a surface he experimented with different materials to attach on and then remove easily without spoiling the hooks. Succeeding in his experiments after about 14 years he patented the discovery in 1955 under the name Velcro. Improved versions of this adhesive tape are so popular and reliable that people use them for even delicate situations like in medical bandages and transporting glass panes.

While considering observation and inference, the story of Karl Jansky, a radio engineer who worked at the Bell labs USA is worth narrating. In 1930, Karl was given the assignment of finding the cause of continued disturbance (interference) in telephone communications at a particular frequency. To investigate Karl constructed a special (linear directional) antenna that helped him identify all the sources of disturbance except one. After several attempts and months of recording the signals and analyzing, he concluded that the disturbing signal was enhanced whenever the antenna was directed towards the center of the galaxy, the Sagittarius constellation. From this, he inferred that the disturbing signal was of cosmic origin and not man-made. This observation eventually gave birth to Radio astronomy now one of the main branches of astronomy that provide us with a novel window to probe the Universe. It includes the study of very high-energy cosmic sources like Quasars, Pulsars, and black holes, as also advanced developments in the science of Cosmology. Unlike this, the acceptance of the Big bang theory for cosmology against the steady-state theory in 1965 with the discovery of the cosmic background radiation is an unusual example of the acceptance of a theory with a single-point observation of the predicted curve (black body radiation). Here the deduction of its significance came from the theorists who interpreted the observation as they were familiar with the prediction that had come a decade and a half earlier (1948 by Gamow et al). Though the complete uniformity of the radiation (isotropic) gave the impetus to discard the competing theory (steady-state) it is the observation of the anisotropies (as predicted by Zeldovich) in the eighties that confirmed the Big bang theory of the Universe as these anisotropies helped explain the structure formation in the Universe to a better fitting. This is an important feature to take home, the significance of repeated observations despite the general beliefs about the veracity of any scientific theory.

Gregor Mendel and his experiments with 'peas' is one of the cases where a rigorous continued effort by one individual found success much later and got established. Mendel acquired expertise in artificial fertilization in plants and carried out numerous experiments of cross-fertilization to determine the process of transfer of 'traits' among the hybrid plants. Mendel noticed that crossbred offspring showed odd traits when the parents had contrasting traits. Mendel took about 8 years of rigorous and meticulous observation and statistical analysis of data from almost 30,000 plants to deduce the laws of genetic inheritance (Pelligrini, 2014). It took

30 years for his model to be accepted, and his work is now considered as a basis for the science of genetics. In one of his publications Mendel mentioned, "It indeed required some courage to undertake such far-reaching labors. It appears however to be the only way in which we can finally reach the solution of a problem which is of great importance in the evolution of organic forms" (Spangenberg & Moser, 1994). Mendel's work is one of the perfect examples of the use of scientific methodology and rigor to provide new knowledge.

Among the more recent (later part of the twentieth century and the current century) discoveries that enhanced the knowledge about our Universe one could consider, observations of the anisotropies in the microwave background radiation, the discovery of the binary pulsar (1974), the cosmic acceleration (1995), and the detection of gravitational waves (2015) on earth, as highly significant.

The discovery of anisotropies in the microwave background (predicted by Zeldovich and Sachs and Wolfe) came mainly from the advanced satellite-based space technology, and helped in explaining the structure (galaxies and clusters) formation in the Universe which generates different perturbations in the distribution of the background radiation (Huterer & Shafer, 2018; Durrer, 2001; Ratra & Peebles, 2003). The binary pulsar discovered by Hulse and Taylor (Weisberg et al., 1981a, b; Taylor & Weisberg, 1989), gave the most important handle to analyze the changes in the binary orbits caused due to the emission of gravitational waves which was indeed a forerunner for the project LIGO (Laser Interferometric Gravitational-wave Observatory (Barish & Weiss, 1999)) that finally detected the gravitational wave signals (emitted by colliding black holes far away in space) on earth in 2015 (Abbott et al., 2016). The new observations made in the late 90s of the faraway galaxies (supernova cosmology project and high z supernovae) revealed an accelerated expansion of the Universe, which was confirmed further in the clustering of galaxies. The explanation for this acceleration was attributed to the (ill-famous) Cosmological constant Λ, but subsequently related to the 'dark energy' which introduces negative pressure in Einstein's equations (Huterer & Shafer, 2018; Overbye, 2017). Observational astronomy/cosmology has thus expanded the view of our Universe manyfold, thanks to the efforts of generations of scientists and advances in space technology.

In the field of life sciences, Edward Jenner (late 1700s) is credited with the discovery of vaccination against smallpox by using lesions from workers who milked cows with cowpox, a less virulent form of pox that rendered them immune to smallpox (Gower, 2020). His understanding of the natural world inspired him to create an effective way to block the scourge of his country and a global threat. As can be expected there have been cases of instant success like in the case of Penicillin (1939) and insulin (1922) both of which seemed to have worked effectively and thus did not require continued experiments over long periods.

In the medical sciences, clinical trials play a very important role in leading to discoveries as the data gathered on different groups of patients when analyzed methodically could trace causes for various effects which then can lead to the possible discovery of new drugs and cures. Unlike in the earlier era, where individuals worked sometimes in isolation, with technological innovations and data mining methodologies, most of the experimental researches in science is carried out by groups of scientists hailing from varied but connected areas to study/observe and theorize logically consistent models for all recorded observations. The remarkable developments in computing technologies and techniques of artificial intelligence today provide major support for the scientific community in analyzing and selecting the most fundamental and suitable data for any particular investigation which are very helpful in saving time and efforts.

The process of science involves apart from hypotheses, experiments and observations, analysis, rigorous check on reproducibility, and then the creation of a theoretical formalism (a paradigm) that enables prediction and validation. The process is endless, with each step resulting in an improved understanding. The creative process and its progress are slow and exhausting but eventually lead to new knowledge that helps the society through its applications, in developing new science or technology. The importance of continued observations and deriving/ rectifying the inferences on its basis cannot be emphasized in any better manner than to quote Sir C. V. Raman (1968) (Raman, 1968):

> The progress in science and the facilities provided by its applications as one sees today, has been mainly due to the efforts of a number of individuals, who have devoted all their time, efforts and possessions towards human progress. From the early philosophers, over the last four hundred years, one can see how the tasks of observation and inferring have gone together to sustain and satisfy the creative minds in developing theories and aids for understanding the Nature we are a part of. As technology helps the observer, mathematical logic helps the theorist.

Learning anything requires concentration of mind, and determination to follow through the process of observation, analysis, and inferencing. Science is no exception or more demanding on these aspects. As mentioned in the previous chapters, learning and practicing science is a very devoted and arduous but enjoyable task which the human mind has adopted and been doing for the last five hundred years and will continue to do so for understanding and trying to answer when, where and how of our universe. As Bertrand Russel points out *"The triumphs of science are due to the substitution of observation and inference for authority. Every attempt to revive authority in intellectual matters is a retrograde step. And it is part of the scientific attitude that the pronouncements of science do not claim to be certain, but only to be the most probable on present evidence. One of the great benefits that science confers upon those who understand its spirit is that it enables them to live without the delusive support of subjective certainty."* (Russel, 1953).

Creative minds are always active and when confronted with impediments in their work often turn to some other avenue of creativity like music, painting, poetry, or some form of art. Many great scientists have shown remarkable flair in different forms of human activities like music, poetry, painting, cooking, etc., suggesting that creative minds tend to show creativity in everything they not only do but even propagate the importance of such activities. Einstein was known to have had good skill in playing violin and it is amusingly said that some people regarded him more as a musician than as a scientist. It seems he rarely left home without his music, and it inspired him as he developed some of the most elegant theories in science. Life without playing music is inconceivable for me, "he seems to have declared". *"I live my daydreams in music. I see my life in terms of music. I get most joy in life out of music*" (MW17). Richard Feynman was supposed to play Bongos to relax while working hard on scientific problems. Werner Heisenberg played Piano. Galileo, credited with developing the correct and useful scientific methodology was supposed to have played flute, written poetry, and painted with water colors. He used his paintings to illustrate his concepts in astronomy. C. V. Raman turned to music, particularly to the sounds of musical instruments, the Tabla and Mridangam which are the main percussion instruments of Indian music. His interest in music followed his inspiration for the science of light as it was for Lord Rayleigh who seems to have conducted experiments on vibrations of bells. Raman himself was keenly interested in understanding the distinct sounds of Mridangam and the physics associated with it (Mulki, 2018). Such traits of brilliant scientists go to establish the fact that human creativity is such that in learning and practicing science, one observes nature and the harmony that manifests itself in varied forms and thus expresses itself in different modes for appreciating the observed. Amongst these various diversions, music and painting appear to be the most influential. In ancient India, traditional knowledge depicting many of the mathematical truths was expressed in terms of Sanskrit verses which helped the growth of both literature and science.

As observation leads to reasoning, it is no surprise that creativity is indeed essential to hypothesize and seek clarification either through referencing earlier mentioned facts or to look for new phenomena that may come out of the newly made observations. As recounted above starting from the early stalwarts like Tycho Brahe, Galileo to Röntgen and Bell with several in between spent almost all their life looking out to different results of their observations and build step by step the ladder of scientific knowledge.

References

Abbott, B. P., et al. (2016). (LIGO Scientific Collaboration and Virgo Collaboration). Properties of the binary black hole merger GW150914. *Physical Review Letters, 116*(24), 241102. arXiv:1602.03840.

Barish, B. C., & Weiss, R. (1999). LIGO and the detection of gravitational waves. *Physics Today, 52*(10), 44.

Durrer, R. (2001). arXiv preprint astro-ph/0109522. arxiv.org

Gower, O. (2020). https://wellcomecollection.org/articles/ (welcome collection)

Huterer, D, & Shafer, D. L. (2018). arXive:1709.01091v2 (5.Feb)

Mulki, M. (2018). *C. V. Raman's work on Indian music.* https://www.livehistoryindia.com/story/cover-story/c-v-ramans-work-on-indian-music/

Overbye, D. (2017). Cosmos controversy: The Universe is expanding, but how fast? *The New York Times.* Retrieved February 21, 2017.

Patrick, C. (1955). *What is creative thinking.* Philosophical Library.

Pellegrini, L., & Tasciotti, L. (2014). Crop diversification, dietary diversity and agricultural income: Empirical evidence from eight developing countries. *Canadian Journal of Development Studies / Revue canadienne d'études du développement, 35*(2), 211–227. https://doi.org/10.1080/02255189.2014.898580

Raman, C. V. (1968). Prof. C.V. Raman's lecture delivered on December 22, 1968, on the Foundation Stone-laying ceremony of the Community Science Center, Ahmedabad.

Russel, B. (1953). *The impact of Science on society*, Routledge edition, 1961.

Spangenburg, R., & Moser, D. K. (1994). *The history of science* (in the Nineteenth Century), (91–95)

Taylor, J. H., & Weisberg, J. M. (1989). *Ap.J., 345*, 434.

Weisberg, J. M., Taylor, J. H., & Fowler, L. A. (1981a). *Scientific American. JSTOR.*

Weisberg, J. M., Taylor, J. H., & Fowler, L. A. (1981b). Gravitational waves from an orbiting pulsar. *Scientific American, 245*(4), 74–82. https://doi.org/10.1038/scientificamerican1081-74

Chapter 9
Science as a Profession

Introduction

The question of whether Science can be a profession, often burden the minds of both students and their parents. It is rather unfortunate that science as a profession does not command the same prestige and respect as the so-called professional courses like medicine, engineering, law, and the managerial cum administrative professions. Of course, the situation is much better today as compared to a couple of decades ago, but still, science as a professional choice needs far more encouragement from society. Having talked about what science is, how it should be practiced and the intellectual satisfaction that it can give, one looks at the present scenario about science as a profession. To begin with, practicing science is not a 9 to 5 job! It is more for the intellectual satisfaction that one finds while being engaged with a problem and finally finds an acceptable solution that matters. It is the only profession where one gets paid for doing what he/she loves to do and moves a small step ahead in understanding Nature. It may so happen that while doing so, one discovers (or innovates) something new that becomes applicable to society. There are several avenues for pursuing science as a profession. One must say that over the last two decades the financial remuneration for science professionals even in India is far better than what it used to be a few decades ago and chances of getting a good position depend mostly on one's merit, dedication and contribution.

One of the main purposes of higher education for any individual is to seek career opportunities for a position of prestige and wealth. How good then is scientific research as a career? The answer lies in the fact that scientific research is perhaps the most fascinating and rewarding of all professions, as one is paid reasonably to further his/her intellectual pursuit for increasing the knowledge about the Universe and contribute towards the benefit of mankind as well as have intellectual satisfaction.

Research in science deals with understanding the laws of nature in their various manifestations. In its pristine form, research provides a way to elucidate the

© The Author(s), under exclusive license to Springer Nature Switzerland AG 2022
A. R. Prasanna, *How to Learn and Practice Science*,
https://doi.org/10.1007/978-3-031-14514-8_9

fundamental reasons of how and why of the way the Universe and its sub-components are and work. Science seeks to determine the fundamental laws, from which one could explain all the events and processes that occur in the Universe. Unlike in philosophy and religion, the keywords in science are hypothesis, experience, experimentation, observation, reproducibility, and verification. Careful, logical, and deep thinking would eventually lead to a logical explanation of every aspect of nature that one seeks to understand. Limitation of the current knowledge about the Universe implies that science cannot claim to explain everything, but its use certainly does provide ways to find some answers with logical reasons. In science, there is always a cause and effect relationship and understanding these relationships in a logical and reproducible manner is its goal. Careers in science are genuinely some of the most exciting, interesting, and significant amongst the various possible careers. The cartoonish character of a scientist as many times depicted- a long-haired and bearded absent-minded individual mostly antisocial is not true. Of course, often a working professional scientist spends a lot more time either in his/her lab or a library even foregoing sleep and regular meals in order not to disturb the thought processes that accompany the research efforts. The often remarked cliché 'sky is the limit' for endeavors, in general, need not apply in a scientist's job as in this profession one is always looking to cross all limits to gain new knowledge. It is true that to progress in the scientific research path, perhaps one needs to be very intelligent, hard-working, logical, and ambitious than in many other areas of career paths.

A research scientist's profession is not merely a job, it is a career where one gets paid for following one's passion and doing so without limitations. The joy that one derives from solving a nagging question is unparalleled. The satisfaction and fulfilment that one attains in understanding and explaining a phenomenon are all that matters to find professional satisfaction. Of course, there are some aspects which can cause worry. As permanent job opportunities for a research-oriented career are limited, one may have to spend a few years in a temporary position of a post-doctoral position which most of the Ph.D. holders in pure sciences do. Along with this, they may be required to travel from one location to another as the post-doctoral fellowships often come only for a limited period. This automatically tells on the person's starting a family as there is never a guarantee of both the person and the spouse getting even fellowships at the same place for the same period. Once a person is willing to go through such hassles the reward of intellectual satisfaction for researching a topic of one's interest overcomes all other disadvantages. As mentioned earlier, in most of the world, financial success and the associated material benefits are considered the most important requirement in life. However, as often stated by great personalities, for wholesome contentment it is the intellectual achievement that surpasses all other forms of professional achievements and wealth. The famous Sanskrit poet Bhartruhari (450–500 AD) has said in his 'Niti Shataka' (hundred verses on morality) praising 'learning and scholarship',

Na chora haryam, nacha Raja haryam
    Na brahtru bhajyam, na cha bharakari.
    Vyayekretae cha Vardhate eva nityam,
    Vidhyadhanam Sarvadhana Pradhanam.

Neither a thief can steal, nor a King can take away; brothers don't ask for division and it is not heavy to carry. As you spend, it increases, Vidya (learning) is the most precious of all forms of wealth.

Getting back to the issue on hand, scientific research broadly comprises of (a) fundamental research and, (b) applied or technological research.

Basic research is mostly sponsored by the governments, and occasionally by the industry. This investment from the state is needed as the returns from scientific enterprise in any undertaking may not be necessarily immediate.

For anyone seeking a career of intellectual achievement and enterprise, with full academic freedom to select one's field of work, a research career in science would be the profession of choice. One would have heard or known about many young persons who choose otherwise due to parental and peer pressure and later regret all their life. Getting an advanced degree in science does not restrict any individual to a life confined to a laboratory as there are different avenues where he/she can use the expertise obtained in different ways while doing science. Learning science in a systematic way and drawing conclusions based on evidence helps in any job including managerial positions. The scientific way of analyzing a given data and arriving at logically consistent solutions to problems is always a sure way to succeed irrespective of the nature of the problem. (It may be useful to take a look at the following website in this connection. [1]). A profession needs to be an engagement of the individual to earn a decent living standard and also contribute towards the welfare of the society he/she lives in. The scientific profession certainly helps in being intellectually active and looking for avenues for the growth of the community to acquire a scientific temper.

Let me at this stage say a few words regarding research in pure sciences which normally is the priority of most youngsters looking for science as a profession. Physics considered the most basic of sciences, deals with the understanding of the fundamental laws that govern the Universe, both at macroscopic (large scale, classical) and microscopic (small scale, quantum) levels. At the macroscopic level, the basic interactions are the gravitational and electromagnetic ones whereas at the microscopic level one has the strong and weak interactions (nuclear forces) along with the electromagnetic ones. Over the last 300 years, a great deal about these forces of nature has been learned both at the macroscopic (classical) and microscopic (quantum) regimes. For the last hundred years, a key problem that has occupied the efforts of several physicists has been the unification of all fundamental interactions. While there exist, unified models of the electromagnetic, the weak, and the strong interactions which indeed led to very important developments and discoveries regarding the fundamental particles of nature, so far attempts to unify gravitational interaction with these three have not been successful. The main hurdle in achieving a complete unification of the fundamental forces has been the problem of 'Quantizing Gravity' i.e. understanding the Gravitational interaction as a space-time phenomenon in the micro-world. Gravity in its macroscopic manifestation though has a complete description through the works of Newton and Einstein, a fuller understanding of the early universe, the structure (Galaxies and Clusters) formation, and the characteristics of high energy cosmic sources, and a complete understanding of black hole physics have yet unresolved questions. The most discussed topics in this area concern cosmological issues regarding the so-called 'dark matter and the 'dark energy. These concepts were brought into focus in the context of attempts to explain observational features like the rotational curves of galaxies (explained by the missing mass) and the accelerated expansion of the Universe (interpreted through high redshift supernovae) in astrophysics. Unfortunately, it has not been possible so far to identify any of the theoretically proposed features of the fundamental particle physics to either of these 'dark' objects. At one stage people had high hopes of answering some of these aspects of high energy physics with the idea of 'super symmetry' (that invoked supersymmetric partners to all the known particles) but the complete lack of evidence on any of the 'sparticles' so far seems to have decreased attention from the notion of supersymmetry and its outcome. It is indeed a great privilege for today's youngsters to look out for newer findings from the recently added facility the **James Webb telescope**. Due to its unique position in space and the advanced technology it carries, astronomy and cosmology will grow richer and more exciting.

Another important development over the last three decades is the discussion of 'quantum entanglement as a driving force in the area of quantum optics. As the related questions involve information transmission and in turn to quantum computing, the applicability of this basic research should be very enticing. Experimental realization of 'Bose-Einstein Condensates', has added incentive to this area of research particularly given the extraordinary probability of finding a way for quantum computing through BECs. As is clear, these issues of very fundamental nature like deciphering the vast unending Universe and attempts to understand their intricate explanation require a deeper knowledge of both physics and mathematics.

On the life sciences front that includes chemistry, the challenge lies in the areas of genetics and understanding of life and its many manifestations. It is interesting to note that despite certain old debates between Darwin's view of evolution and the old beliefs concerning life evolution, modern understandings from the science of molecular genetics the sequences of mutations in the genomes perfectly seem to match the evolutionary history as deduced by Darwin. This fact should increase the scientific approach to understanding what life is, a question that can give a new and firm basis to the investigations in chemistry and biology of life. Particularly it is necessary to understand how the existing life forms emerged from abiotic chemistry. Laboratory experiments have succeeded in creating amino acids that are basic to life and going further to understand the creation of life itself has been a challenging issue that can be a very interesting and absorbing enterprise. On the applied front of these studies, genetic engineering and biotechnology have quite a lot to offer, as one could look for beneficiary fallouts that are useful in medicine and agriculture. Environmental chemistry relates a host of interesting problems relating to each other and has a direct impact on societal needs and responsibilities. Whereas pollution control has been talked about at various levels efforts to control by taking actions on industrial waste disposal has not been encouraging. It may be a very useful topic for chemists to work on problems relating to studying the possible changes one could make through different chemical processes of the waste material which may help in coverting waste into something useful.

Astrobiology is today a very interesting and sought-after line for research. The discovery of long-chain molecules in the space environment has brought in a lot of interest in trying to understand the so-called 'diffuse interstellar bands and consequently a closer link between biologists and astronomers which indeed is a very healthy unity for science particularly in the context of exoplanets research. As the interest in space technology is getting widespread there seems to be increasing investment in space science and consequently, astrobiology is getting good attention both from researchers in physical as well as life science disciplines. The astronomical findings of the exoplanets have increased the interest in the search for extra-terrestrial life and intelligence and research in this area is quite a complex combination of all the basic disciplines. In this context, it becomes important and necessary to look out for the energy requirements for repeated space programs and this takes one to the technological innovations that may have to be applied for the utilization of solar energy which is the only abundant source outside the Earth. As mentioned earlier the relationship between science and technology itself can be a good research topic that might have useful offshoots that could help society in general and material scientists in particular. In this context, it is useful to bring up one of the recent (during the last three decades) interests of research- the nano-science and the nano-technology. These areas of research are very adaptable particularly for those who enjoy practicing science thematically using knowledge from different areas of physics, chemistry, and sometimes biology and mathematics too. It could bring several difficulties in arriving at conclusions as the applicability would differ depending upon the exhibited property. As reviewed by Khan et al. (2019), the importance of nano-particles was realized when it was found that their size can influence various physio-chemical properties like their optical features. Dreaden et al. (2012) have reported that these particles show characteristic colors and properties with the variation of their shape and size which seem to depend on their absorption property. It is suggested that the optical and electronic properties of these are interdependent like in the case of noble metals. Nano-particles of noble metals having size-dependent optical properties exhibit a strong UV–visible extinction band that is not present in the spectrum of the bulk metal. Research in these aspects could be very stimulating as apart from being intrinsically rich in the interlinked understanding they can also have several applications in various technological applications. The nano-particles are not simple molecules but composed of three different layers consisting of (a) the surface layer, which may be functionalized with a variety of small molecules, metal ions, surfactants (substances that reduce the surface tension of a liquid when dissolved like sodium stearate in soaps) and polymers. (b) The shell layer, which is a chemically different material from the core in all aspects, and (c) The core, which is essentially the central portion of the nano-particle and is usually referred to as the nano-particle itself (Shin et al., 2016). The importance of nano-particles in the development of new nano-devices that are useful in the physical, biological, medical, and pharmaceutical industries including space technology can open up immense possibilities for non-invasive body examinations as well as in determining the optimal dosage that a patient may need of a particular drug.

Having mentioned these various possibilities for research in nano-science and technology, it should be obligatory to briefly draw attention towards the toxicity factor associated with nano-particles (Bahadar et al., 2016). It is quite clear that the nano-particles enter wherever they want surreptitiously through various human activities and could become potentially lethal factors by inducing adverse cellular toxic and harmful effects, which could be unusual in micron-sized counterparts. Studies seem to have illustrated that nano-particles can enter organisms during ingestion or inhalation and can translocate within the body to various organs and tissues where the nano-particles can exert the reactivity that can bring toxicological effects. Thus it is very significant to understand the importance of nano-particles and their science in all aspects.

These are very few examples of pursuing and practicing science to make it a profession of interest and enterprise. The important message one ought to take while choosing science as a profession is the fact that in the span of a few hundred years science has successfully explained many of the observed natural macro phenomena, like thunder to floods, droughts to forest fire, eclipses to comets and the motion of planets and stars and also trying to understand more significant micro features concerning all aspects of life and the cosmos through the study of micro and nano features of the matter. Both in physical and life sciences numerous exciting areas could give personal satisfaction as well as help develop new applications of value to humanity.

Coming to the important branch of pure mathematics, which according to general feeling is completely theoretical, the professional possibility is in getting a professorship or if interested in the applied aspects specialize in computer science and its usage in network analysis, optimization, and finance (actuarial mathematics). Some of the current topics of investigation popular in the area of pure mathematics deal with the Number theory, Algebraic geometry, Differential geometry, Representation theory, and Mathematical biology. Further, there has also been a lot of interest particularly from the community of computer scientists, on topics like Discrete mathematics, Game theory, Data mining, and Quantum computing. The interaction of mathematics with almost all other disciplines of learning has been on the increase.

Until about a few decades ago, some topics (mathematics, theoretical physics, astronomy) offered possibly only academic openings like teaching jobs, while some others (chemistry, biology, and its associated branches) had possible openings in the industry (pharmaceutical and medical) too. However, things have changed of late, as new understandings of the basic ideas from overlapping features of matter and its manifestations are becoming clearer in various forms. This indeed has created very many new opportunities for graduates to migrate from one subject area to another looking for applications that are interesting and have the possibility of being useful. This possibility has increased manyfold because of the applications of computer programs in data analysis which is a requirement in almost every field of research and development. The most interesting subjects in interrelated sciences are biophysics, mathematical biology, chemical kinetics, micro, and molecular biology and genetics, neuro-psychology, astrobiology, robotics, and biomedical applications, and similar combinations. These topics are inherently deep and offer varied possible research topics that are mentally stimulating and also show possible applicability.

There have been several engineering graduates who have turned their attention towards pure science and the advances in nanotechnology and associated electronics have opened a host of opportunities for technically minded students with an interest in basic science to pursue pure research.

Research in the areas of geophysical and geochemical studies of both the Earth's interior and atmosphere, along with oceanography would help in understanding the climate and weather systems in the world. Closely related are topics of environment and ecology which would involve several aspects of physics, chemistry, and biology and some amount of mathematics too.

Apart from clearer understanding and inventing new technologies what is most important in a research career is the resonance of the researcher with his/her topic of investigation. This is best done when one thinks of a research topic for oneself that enthuses and stimulates one's mind harmoniously and encourages one to develop the background and techniques required for pursuing the goal. Such an attitude helps one to go deeper without worrying about immediate returns.

While both fundamental and applied research has seen an increased interest from students and opportunities also are increasing, it is very important to keep in mind the ethical issues that may appear while dealing with topics like genetic engineering, stem-cell research, chemical and biological weapons, and innovations in robotics which in the wrong hands can lead to despair and destruction. This is of particularly great importance in the context of research in the area of new genetic technology called CRISPR (system consisting of Clustered Regularly Interspaced Short Palindromic Repeats) which is supposed to offer more control over DNA changes which could open up methods for altering genomes. This can lead to unintended effects as they are prone to errors, though with the safe and careful application may be found useful for gene therapy as well as help develop models of diseases (Ledford, 2019). Reports exist that such studies in genome editing have caused extensive mutations which presently are not detectable with existing DNA tests (Dockrill, 2018). Such investigations do certainly require strict supervision and control as otherwise one may be led to the legendary 'Frankenstein monster'.

In the last two years interest in the disciplines of virology, immunology, and associated fields have increased and are considered as being very important due to the COVID pandemic. This certainly needs young minds getting into research in these areas that would help the medical practitioners as well as the pharmaceutical industry to a great extent and help the societal needs.

Scientific research is a noble profession when practiced with care and responsibility, both for the advancement of knowledge about ourselves and of our Universe. Of course, no one can guarantee that all possible outcomes of new investigations, particularly in certain delicate areas like genetics-related research, yield positive results unless done under strict and conscientious supervision. A career in science can be most lucrative when one takes it up as an intellectual challenge rather than a routine 9 to 5 job. The excitement of science is derived from the fun of exploring the unknown through imaginative hypothesis building and then deriving possible solutions through axioms and logic. The quest to understand the Universe in all its manifestations, as well as communicating it to society meaningfully can be and is a deeply satisfying intellectual endeavor.[1]

References

Bahadar, H., Maqbool, F., Niaz, K., & Abdollahi, M. (2016). Toxicity of nanoparticles and an overview of current experimental models. *Iranian Biomedical Journal, 20*(1), 1–11. https://doi.org/10.7508/ibj.2016.01.001

Dockrill, P. (2018). Nature BioTechnology. https://www.sciencealert.com/crispr-editin

Dreaden, E. C., Alkilany, A. M., Huang, X., Murphy, C. J., & El-Sayed, M. A. (2012). The golden age: Gold nanoparticles for biomedicine. *Chemical Society Reviews, 41*, 2740–2779.

Khan, I., Saeed, K., & Khan, I. (2019). Nanoparticles: Properties, applications and toxicities. *Arabian Journal of Chemistry, 12*, 908–931.

Ledford, H. (2019). *Nature, 574*, 464–465. https://doi.org/10.1038/d41586-019-03164-5

Shin, W.-K., Cho, J., Kannan, A. G., Lee, Y.-S., & Kim, D.-W. (2016). Cross-linked composite gel polymer electrolyte using mesoporousmethacrylate-functionalized SiO_2 nanoparticles for lithium-ion polymer batteries. *Scientific Reports, 6*, 26332.

---

[1] See https://www.ncl.ac.uk/careers-occupations/science/out-in-the-field/

Chapter 10
Science and Society

Introduction

As Bertrand Russel (1953) says *"the effects of science are of very different kinds. Most important of them being of the intellectual effects which help in dispelling traditional beliefs. Apart from its effects on industry and war there are profound effects in social organisations mainly due to the new technology that science has brought in. This has consequent influence on the political system too. As a result of the new control over the environment due to scientific understanding there is a new philosophy involving a changed concept of the man's place in the Universe"*.

In one quote we find several issues all of great concern to the humans and the society they define. In the earlier sections science as an intellectual pursuit has been discussed with all its manifestations. As quoted often, science means knowledge. A glance at the history of science reveals that knowledge advancement and understanding nature in its various forms have only been possible due to the untiring and dedicated efforts of men and women analyzing what they observed and what they experienced. Early Greeks like Plato and Aristotle looked for patterns in the behavior of things both living and non-living. It was conveyed that nothing should be taken for granted and whatever knowledge one had from the earlier generations should only act as stepping stones for new adventures. This brought in the culture of "Research" which by the very name indicates searching again and again for knowledge based on already familiar phenomena to include new observations and views that follow. As expressed in the earlier chapters with examples this involves different stages of preparation, incubation, illumination, and verification. Scientific research is thus a continuous process with a firm beginning but never-ending.

© The Author(s), under exclusive license to Springer Nature Switzerland AG 2022
A. R. Prasanna, *How to Learn and Practice Science*,
https://doi.org/10.1007/978-3-031-14514-8_10

Public Perception of Research

Scientific research what is it? Being a practicing mathematical physicist, I have often been asked by my non-science acquaintances; "What? research in mathematics, is there still something to do in mathematics?' Perhaps one cannot blame them as it is the failure on the part of 'pure scientists', who sit in their ivory towers thinking that they have no time to interact with the society or spending time to tell in simple terms as to what they are studying and why. An interesting anecdote that comes to mind is as follows. Way back in 1980, there was a meeting at the International Center for Theoretical Physics, Trieste, and Professor Arthur Koestler, a distinguished scientist was addressing a gathering of administrators. He recalled an incident when a minister of the cabinet was visiting the University of Paris in Saclay. Apparently while going around the various facilities including the laboratories, he saw a young man quite well dressed whom he asked as to what research he was doing. The youngster replied with a little flourish, 'high energy physics'. The minister just nodded and moved forward till he saw a young lady and he repeated his question. She responded 'low energy physics'. The minister beaming went to her and patted her saying *'Excellent, that is how I want our young ladies to be 'honest and modest'.* It is a clear example of the lack of understanding of the concept of scientific research even to those who are supposed to govern and administer the country's develop-mental activities.

Another story that was told to me by one of my uncles goes like this. Two businessmen met at a club and were talking about new investments. Someone suggested that chemistry research would yield lots of profit citing the example of Nobel and the dynamite. They agreed with the idea and soon set up a beautiful laboratory and engaged a chemist. The next day when they met for lunch one of them said 'let's go and check what the guy has discovered', the other responded— *'No let's wait till tea time that will give him a few more hours before we check'.*

The above anecdotes may appear exaggerated but the public perception of scientific research, unfortunately, borders on such misgivings. To appreciate the notion of research, one should have a 'scientific temper, which mostly is absent among the masses which is called the society. Having a scientific temper provides one a framework of thinking and patience to understand which are very essential to gain knowledge in any form.

It is indeed a pity that the public seems to get its scientific inputs mainly from the media which describes mostly about a new gadget or a new utility product, catego-rizing them as technical advancements. While there is nothing wrong with this, it really may not convey the science underlying the technology. Here again, the fault is with some scientists (?) themselves who run to the media with their findings for getting popular even before a proper verification of their finding. Of late many serious science practitioners have tried to make the public aware of new develop-ments in non-technical terms but still the numbers of such interventions are far less to create a proper understanding of the rigors of science. By and large, if the misgivings of the society about scientific research have to be alleviated then the scientists ought to take courage and learn to communicate to the common man as to what they are doing and why is it important. Aspects of applied science indeed catch the fancy of the public far more easily than of fundamental science. But one should realize that today's science is tomorrow's technology. If we are enjoying life today with several amenities for day-to-day living, it is essentially due to the insight and dedicated hard work of a galaxy of scientists like Newton, Marie Curie, Faraday, Rutherford, Bohr, Einstein, and many of their distinguished contemporaries and followers for whom research was and is a passion and its culmination in discoveries or inventions only incidental. Luckily in their time, they were not asked to write a 'project report' as to what they would discover, and of course, they never had to apply for grants from the government neither had to rush to get patents.

Change in the Scenario

Times have indeed changed and today scientific research is a profession that needs full support of the society. To appreciate the real needs of scientists, society should necessarily listen and appreciate the views expressed in developing and leading a life of inquisitiveness with a desire to understand. To support scientific research the public should try and spread the 'expressed views with logical understanding' among themselves, thereby making efforts to change all 'false beliefs, superstitions and irrational practices which engulf the society. The lack of mass communication in science can sometimes lead to harmful situations regarding the health of the people as pointed out by Prof Saroj Misra in a brief message on mass communication and public health. As she points out the main stream media does not seem to publicize causes and preventive measures in general. In the last two and a half years because of the pandemic gripping the entire world population, there are some efforts to change the situation but what one hears about the refusals to take vaccination by some even in the advanced country like US seems to be a real problem. It is very unfortunate that political leanings seem to take upper hand even in such a situation and those who practice politics irrespective of where they are from do not give importance to mass awareness of the principles of science. Once again fortunately there are a few industrious and intelligent individuals who try to pioneer the cause of mass com- munication of scientific practices and its values which will hopefully rectify the ill effects of mass ignorance. In the context of science communication and public outreach there have been several different opinions regarding to who all and at what level scientific information needs to be conveyed in the public domain. As an example, considering the situation in the US (as processed information is available) though the National Science Foundation surveys seem to report that almost 90% of US adults claim to be interested in news about science and technology, according to Miller's estimation (1986) only about 20% of the American adults are attentive to science policy-a group which tends to be younger, better educated, and likely to have college level science education. Continuing on, it is pointed out that about 40% can be characterised as 'science interested'-a group who do have a relatively high interest in science and technology but lack functional understanding of the process of science. Compared to the earlier mentioned group this consists of slightly older, somewhat less educated, and less likely to have had college-level science courses. It is pointed out that while many people profess interest in science, a high percentage of even the attentive public do not seem to have scientific literacy as reported by the National Science Board (2000) as they could not answer simple questions related to facts of science. While such findings do raise questions about the content of public understanding of science which could indeed be true in most of the countries (developed and developing) the traditional view holds that all citizens involved in initiating/promoting public scientific policy, should have basic scientific literacy. On the other hand, those of the citizenry who are policymakers (for example take decisions regarding budgetary allotments) should have a sufficiently high level of scientific knowledge (not just information) that will help both the leaders and the science practitioners/professionals. To ensure such a possibility the general public everywhere should be a target of science communication (Miller, 1986; Yankelovich, 1982; Wiegold, 2001). As science is not a visible occupation (briefly pointed out in the last chapter) the attitude people hold toward science is complex. While there may be some who feel that 'technology will find a way to solve the problems of the society, a lesser number agree to the statement 'everything has a scientific explanation'. Another survey by the National Science Board (2000) seems to have shown that while about 70% of the respondents seem to agree that 'science is the best source of reliable knowledge about the world', about 40% think that 'technology has become dangerous and unmanageable. When one takes into account the media coverage for communicating 'science news', the 'risk factor' seems to play a large part in drawing the public attention. One of the best examples in this context is that of reports on 'Chernobyl accident of the nuclear reactor (1986)' which impacted the food and other imports from Eastern Europe to a large extent as the general public feared the risk of radiation effects (more from heresy than actual knowledge of facts). The influence was so challenging that the then head of state (USSR) President Gorbachev seems to have said 'Chernobyl accident was a more important factor in the fall of the Soviet Union than Peristorika' (Chernobyl Accident, 1986). As points out the literature on science communication while often portrays the reader as relatively passive and uninvolved, those who read about risk factors are often active. After the unfortunate accident referred to above, any news on the possibility of establishing a nuclear power reactor (mainly for supplying power) anywhere in the world was received with quite a massive public reaction (opposition) particularly at and near the proposed sites.

Einstein in his remarks (1934) on 'society and personality' says "The individual is what he is and has the significance that he has not so much in virtue of his individuality but rather as a member of a great human community, which directs his material and spiritual existence from the cradle to the grave. A man's value to the community depends primarily on how far his feelings, thoughts, and actions are directed towards promoting the good of his fellows". Continuing (Einstein, 1954), he says, "It can be seen easily that all the valuable achievements, material, spiritual and moral which we receive from the society have been brought about in the course of countless generations by creative individuals, as only the individual can think and thereby create new values for the society".

This synergetic relation between the society and the individual is important to understand by practitioners of science too, as their creative existence requires the full support of the society. To develop and keep it going for future generations, working scientists ought to take some time off to mix with the public at different levels of education and communicate to them about the integral role of science in human development putting their research work in perspective. In return, society should first understand the important role that scientists play in the understanding of nature and the world around us and support with enthusiasm the hard work this community does through increased recognition. As political establishments are involved in the governance of any country and as it is the leaders in the society who manage to occupy positions of power one can easily see why it is important for the society to develop a scientific temper. In the modern world particularly in the last couple of decades, the tension in world affairs is on the rise and this has led to the increase of so-called defence budgets of Nations to have supremacy in warfare. The inputs from technology in this context are very well known. It is indeed a pity that this situation has led to spending more on what is called applied research as against spending on fundamental research. It is a known fact that several leading nations spend more resources on having an upper hand in defence equipment in the name of national security. This should not end up in the development of weapons of mass destruction by physical, chemical, or biological means. Here is where a meaningful understanding of societal needs and scientific research would help in setting up science for peace rather than for war technology.

An important issue that needs consideration by the science practitioners is as pointed out by J. Bronowski, the democracy of the society of scientists themselves. "The crux of the ethical problem is to fuse the private and the public needs, recognize thinking above thoughts and the search above the discovery. A true society is sustained by the sense of human dignity". Continuing, he describes that "the society of scientists is simple because it has a directing purpose: to explore the truth. It must encourage the single scientist to be independent and the body of scientists to be tolerant. From these conditions follows a range of values: dissent, freedom of thought and speech, justice, honour, human dignity and self-respect".

As mentioned briefly earlier the development of atomic weapons was an aberration and Einstein who was indirectly alluded to as the proponent is the most unfortunate outcome of the thoughts of a few mostly self-serving people. As Einstein says in his message on 'active pacifism' (1934) (Einstein, 1954) "we must not conceal from ourselves that no improvement in the present depressing situation is possible without a severe struggle, for the handful of those who are really determined to do something is minute in comparison with the mass of the lukewarm and the misguided. In my opinion deliverance can come only from the people themselves". He clarifies that pacifism which does not actively fight against the armaments of nations may remain impotent; only when the conscience and the common sense of the people are awakened one may reach a new stage where people will look back on war as an incomprehensible aberration of their forefathers! The above quotations from Einstein are purely to remember how the discovery of a scientist for the pure purpose of understanding the relativity of motion and explaining the structure of space-time for uniformly moving observers, gave a unique relation between mass and energy. But the way one of its applications was used to kill lots of innocent people was mainly the fault of the society that governed the system. Unfortunately, this misguided conduct has not stopped as governments do spend quite a bit of their defence budget on so-called classified (secret) research mostly in chemistry and biology which many times produce effective results of a negative character. Here again, it is society that should have a voice in controlling such ventures. One wonders whether the so-called national science academies and councils have any say in such matters and if not why is it that the academies are not proactive in maintaining the responsibility of advising and checking the usage of science and technology by the society.

This brings me to another point of view expressed by Bertrand Russel (1953) while discussing the question 'can a scientific society be stable?' His definition of a scientific society implies the degree to which scientific knowledge and technique based on that knowledge affect its daily life, its economics, and its political organization. As he emphasizes the question is whether it is probable or not for a society to be scientific; if it does, it must almost inevitably grow progressively more scientific as the new knowledge accumulates, and if not there may be a gradual decay. The former would show itself in exhaustion and the latter in revolution or unsuccessful war. This point of view has led him to conclude that as knowledge is power equally shared both by the good and the evil unless wisdom increases as much as knowledge the result would be undesirable. Reconsidering his points of view it is very necessary to realize the main causes that could result in instability, the physical, biological and the psychological. The two important physical reasons are (1) excessive agriculture which leads to the exhaustion of the soil and (2) excessive industrialization leading to the exhaustion of natural resources like oil, and mineral ores. The excessive demand for the consumption of power is leading to the utilization of all types of underground resources leading to both excessive carbon footprints and excessive radioactive decay products which in the long run is harmful. The biological reason mainly concentrates on the successful evolution of man using his intelligence and ruling over all other species of the animal world and gaining complete supremacy of even the number. As one knows the animal world has been shrinking and except for the domesticated ones that man uses for his benefit many species have become extinct. Though on the one hand discoveries in medical sciences have reduced the possibility of infant deaths, in many parts of the world children die due to malnutrition, as the non-availability of staple food has increased. This increases the difference and distance between the rich and the poor which can truly get to the stage of insecurity which on a large scale may lead to war and destruction.

To conclude on a positive note, it is essential that we as a society should take every possible step to ensure that every generation develops good scientific temper and thinking capacity to separate right and wrong. Very often monetary concern alone seems to override all other aspects and the wants often take over the real needs. As a member of the society once one learns to consider the needs of all and be responsible to understand requirements of future generations before wasting all natural resources the humanity will grow harmoniously. Science is a very important input to realise this and Society has to participate fully for this realisation.

References

Chernobyl Accident. (1986). https://world-nuclear.org/information-library/safety-and-security/safety-of-plants/chernobyl-accident.aspx

Einstein, A. (1954). *Ideas and opinions*. Rupa and co.

Miller, J. D. (1986). Reaching the attentive and interested public for science. In S. M. Friedman, S. Dunwoody, & C. L. Rogers (Eds.), *Scientists and journalists: Reporting science as news* (pp. 55–69). Free Press.

Wiegold, M. (2001). *Science Communication, 23*(2), 164–193.

Yankelovich, D. (1982). Changing public attitudes to science and the quality of life: Edited excerpts from a seminar. *Science, Technology, & Human Values, 7*, 23–29.

Chapter 11
Epilogue

Learning and practicing in any discipline are individual efforts. Whatever be the field pursued the motivation one has and the discipline with which he/she pursues the goal determines the rate of success. Science is no exception but is more demanding as it needs building up one's acquired knowledge from experience. The experience is gained through observation and or by experimenting. Often when young ones ask a question as to the reasons for understanding a happening many try to ignore or disapprove. The ability to question is the trait one should develop as the questions motivate a person to self-realize the truths about nature.

Science is the outcome of the work and efforts of generations of individuals and groups with sustained curiosity and devoted perseverance to understand the world they lived in and observed around. Continuing the good work is the responsibility of the new generations and supporting the efforts is expected from the society. Learning and practicing are the main initiatives that require thoughtful consideration and realization. Science is now a part of our living as one can see it easily in most of our daily actions and reactions. The cause-effect relationship gives one a lot of motivation for the thinking-a gift that humans have from nature. It is by no means suggested that other life forms cannot think. It may indeed be an interesting topic to investigate the manifestation of thinking in living beings outside humans.

The fact that there is a definite methodology for practicing science developed over a couple of centuries can be considered as a true gift from our ancestors. As the situations change one can adopt acceptable modifications but one should adhere to the established checks and bounds for the veracity of new ideas. As the language of science communication has to be discernible universally the terminologies used must have general validity and understanding. In this context, it has been discussed that the language of mathematics as adopted by Galileo, Newton, and other founding fathers should be made accessible at different levels appropriately according to age and experience. Here for the younger generations, the ideal way would be the thematic presentation of facts citing examples from what they see, touch, or hear. This would link the topics in such a way that one need not make subdivisions like physics, chemistry, biology, etc. but emphasize the characterization of properties

A. R. Prasanna, *How to Learn and Practice Science*,
https://doi.org/10.1007/978-3-031-14514-8_11

that link matter at different phases and associated energies as well as the interactions they possess and exhibit.

Science is an activity of creative and imaginative human beings—qualities that are themselves controlled by discipline and self-criticism. Though creativity, which is imagination in action is natural to man, only when it is used to its fullest extent could it result in a discovery or invention! As expressed by J. Bronowski, (1953) *"Science is the creation of concepts and their exploration in the facts. It has no other test of the concept than its empirical truth to fact. Truth is the drive at the center of science; it must have the habit of truth, not as a dogma but as a process"* ...Continuing, he asserts, *"It is a common and cardinal error to suppose, as the nineteenth century supposed, that the facts on which science builds are given to us absolutely, and call for no judgement or interpretation from us. The great discoveries in the physical sciences in the twentieth century begin from a radical denial of this philosophy. We now understand that science is built not on facts but on observations, that observation is not a passive state of reception, but an active relation between the observer and his world; and that science is therefore not a mechanical index of facts but an evolving activity. Truth to fact is the same habit both in Arts and Science and has the same importance for both because facts are the only material from which we can derive a change of mind"*. It is said that Art expresses the Beauty of Nature as it beholds; Science goes a step deeper and unravels the pattern that holds the beholder!

Having seen the positive aspects of scientific investigations while explaining what science is, it is important to also look at some of the unfortunate consequences that have resulted because of the indiscriminate use of high-powered technology based on scientific principles. One of the unfortunate effects is due to radiation resulting from exposure to it (even small doses) for a long time as it happens in the case of those working with radioactive materials. The most glaring example of this is the discoverer of radium-Mme Curie herself. Workers in such labs are indeed provided with badges that keep registering the daily dosage and indicate timely the limits. But many times this caution might be overlooked in competing to achieve some short-term goals which eventually may be very harmful. Even the X-rays while being very useful, sometimes can affect the healthy cells while trying to destroy the cancerous ones. This may also lead to dangerous consequences that one should be aware of. More often than not, there is a tendency to overlook the long-term harm over the short-term gains. Here is the most important ethical issue that one should bear in mind. The immediate advantages to the existing generation may become a reason for the long-sufferings of future generations!

Every generation searches for more vital things and like the ever-changing march of our clothes, the costumes of creation take on many different fabrics. However, all scientific creations need not be new and lead to discoveries. Very often science has to explain already existing patterns in terms of a language that can be clear to all those who are also engaged in finding out the truths of Nature. As our view of the Universe expands with new developments both in understanding the science and with advancing technologies the efforts need to be ever enhanced. So goes the story of learning and practicing **Science**.

Appendix

Solutions to number puzzles

1. Taxi number $1729 = 12^3 + 1^3 = 9^3 + 10^3$
2. Diophantus, Ah, what this tomb holds a marvel! the tomb tells scientifically the measure of his life let it be x. Childhood x/6. A twelfth was added x/12. Wedded time without children x/7. After five years 5, he got a child who lived only for x/2 years. After 4 years he died. So x = x/6 + x/12 + x/7 + 5 + x/2 + 4. Solving for x one gets the age which was 84
3. Let the number of meters be x The amount received should then be 4936 x ps Entry in the register should be (1000y +728) ps, So, one has 4936 x = 1000y + 728. One equation with two unknowns Indeterminate! Diphontaine equation. Dividing the equation by 8, one gets, 617 x – 125 y = 91. Here both **x** and **y** have to be whole numbers and **y** cannot exceed 999 as there are only three unseen digits.

   125 y = 617 x - 91, or y = 5x −1 + (34 -8x)/125 = 5x −1 + 2 z,
   z = (17 − 4x)/125, 17 - 4x = 125 z, x = 4–31 z + (1-z)/4 = 4 -31z + t, or t = (1-z)/4. Thus, z = 1 − 4 t, x = 125 t − 27 and y = 617 t − 134. But $100 \leq y < 1000$. Hence, $100 \leq 617\,t − 134 < 1000$ which says that $t \geq 234/617$ and $t < 1134/617$. So t has to satisfy $1.8 > t > 0.38$, but should be a whole number which means, t = 1. Using this in x and y, one gets x = 98, y = 483. *Hence, 98 meters of cloth sold for Rs 4837.28.*

4. King and the Vazir
   Vazir brings in another elephant and adds to the herd. Now there are **18**. One half for the first, and so he gets **9**. The second, one third and he gets **6**. The third has to get one ninth and that is **2**. After distributing accordingly, Vazir takes his elephant back as 9 + 6 + 2 = 17, the number King had.

© The Author(s), under exclusive license to Springer Nature Switzerland AG 2022
A. R. Prasanna, *How to Learn and Practice Science*,
https://doi.org/10.1007/978-3-031-14514-8

Further Reading

Bhanu Murthy, T. S. (1992). *'Ancient Indian mathematics'-a modern introduction*. Wiley Eastern.

Einstein, A. (1979). *Ideas and opinions*. Crown Publishers (1954), Rupa paperback.

Eicher, D. L. (1968). *Geologic time*. Foundations of Earth science series. Prentice Hall.

Ehrlich, R. (1997). *Why toast lands jelly-side down*. Universities Press.

Farnes, P., & Kass-Simon, G. (Eds.). (1990). *Women of science, righting the record*. Indiana University Press.

Fuller, S. *The struggle for the soul of science; Kuhn vs popper*. Icon Books.

Gardner, M. (1952). *Fads and fallacies in the name of science*. New York American Library.

Glashow, S. L. (1988). *Interactions: A journey through the mind of a particle physicist and the matter of this world*. NY Warner Books.

Graf, F. R. (1964). *Safe and simple electrical experiments*. Dover.

Hoffmann, B. (1959). *The strange story of the quantum*. Dover.

Hunter, J. A. H., & Mandachy, J. S. (1975). *Mathematical diversions*. Dover.

Jungk, R. (1958). *Brighter than a thousand suns, A harvest book*. Library of Congress no 58.

Kamshilov, M. M. (1976). *Evolution of the biosphere*. Mir publications.

Korenberg, A. (1989). *"For the love of enzymes": The Odyssey of a biochemist*. Cambridge Harvard University Press.

Kuhn, T. S. (2012). *The structure of scientific revolution*. University of Chicago Press.

Lederman, L., & Schramm, D. N. (1989). *From quarks to the cosmos*. NY Scientific American Library.

Livio, M. (2013). *Brilliant Blunders: From Darwin to Einstein*. Simon and Shuster paperbacks.

Macqueen, J., & Hanes, T. L. (1981). *'The living world' exploring modern biology*. Prentice Hall of India.

Michelson, A. A. (1927). *Studies in optics*. Phoenix science series, University Chicago Press.

Moore, J. A. (1993). *Science as a way of knowing. The foundations of modern biology*. Cambridge Harvard University Press.

Moore, R. (1961). *The coil of life, story of great discoveries in Lifesciences*. Alfred A Knopf.

Pias, A. (1986). *Inward bound, of matter and forces in the physical world*. Clarendon Press.

Prigogine, I., & Stengers, I. (1984). *Order out of chaos*. Bantam Books.

Rees, M. (1997). *Before the beginning*. Library of Congress no 98.

Srinivasan, K. R. (2005). *One hundred reasons to be a scientist*. ICTP (Trieste) publication. Hindustan Book agency.

Stoller M and Fomin L, Amusing experiments (1986).

Tikhonov, A. N., & Goncharsky, A. V. (1987). *Ill-posed problems in the natural sciences*. Mir Pubs.

Watson, J. D. (1968). *The double helix*. NY Atheneum.

Wheeler, J. A. (1990). *A journey into gravity and Spacetime*. Library of Congress no 31. Scientific American Library.

Zee, A. (1999). *Fearful symmetry. Search for beauty in modern physics*. Princeton University Press.

Printed in the United States
by Baker & Taylor Publisher Services